I0053977

Renewable Energy's *Broken Promises*

Large-Scale Renewable Energy Project Failures and Success of Rooftop Solar

Douglas B Sims, PhD

Printed in the United States of America.

For more information, or to book an event, contact:
dsims@simsassociates.net

Book design by DB Sims
Cover picture: iStock by Gong Hangxu

ISBN – Paperback: 979-8-9919108-2-8
ISBN – eBook: 979-8-9919108-3-5

First Edition: November 2024

Table of Contents

Acknowledgements

I am profoundly grateful to my wife, whose steadfast support, wisdom, and love have been my anchor and inspiration. The journey we've shared over the past 34 years has enriched every chapter of my life and this project. Your encouragement has lifted me through each challenge, and your insights have helped shape my perspective in ways that are reflected on every page.

To our children, thank you for filling our lives with joy and growth, teaching us both the rewards and trials of parenthood. Watching you become who you are has been one of my life's greatest privileges, filling me with pride and offering lessons that influence my work and my worldview.

To our family, thank you for your unwavering support. Your presence has been a source of strength, and your companionship has been invaluable in both my personal journey and in completing this project.

To my friends and colleagues—especially those in renewable energy professions and political science—I extend my deepest gratitude. Your insights, expertise, and perspectives have challenged and enriched my thinking. Engaging in debates and discussions with you has added a depth and authenticity to this book that I could not have achieved alone.

Lastly, to the many individuals I have had the privilege of observing and interacting with professionally, thank you for sharing your experiences. Your stories have provided invaluable insights, adding real-world understanding that resonates throughout these pages.

Forward

I worked in the alternative energy sector on the front end of environmental due diligence, clearing 100 to 500,000-acre sites for mega energy projects. I witnessed firsthand the massive environmental impacts of these projects, and the inflated promises made to taxpayers that these ventures would solve the nation's energy demands. Too often, they fell short, becoming financial burdens on taxpayers while corporations profited from federal subsidies. Although intended to transform the grid, many large-scale projects—massive wind farms and solar installations have struggled to meet their promises.

In the drive to replace fossil fuels, America has heavily invested in large-scale alternative energy, funneling billions of taxpayer dollars into these projects. But in reality, these ventures are often inefficient, environmentally disruptive, and prioritize corporate profits over sustainability. While initially viewed as grid-scale solutions, they revealed considerable issues: vast land requirements disrupt ecosystems and farmland, while distance from urban centers results in costly transmission inefficiencies. Additionally, the intermittent nature of wind and solar farms demands natural gas backups, offsetting their environmental benefits.

By contrast, rooftop solar has emerged as a more effective solution, sidestepping many of the pitfalls that come with large-scale renewables. Cost-effective and environmentally friendly, rooftop solar enables communities to control their energy needs, reducing reliance on corporate-driven mega-projects. Decentralized and built on existing structures, rooftop solar generates power near consumption points, reducing transmission losses and minimizing land use conflicts.

The push for alternative energy in the U.S. began in the 1970s and grew rapidly in the 2000s, with significant policy incentives aimed at reducing dependence on fossil fuels. Projects like the Ivanpah Solar Electric Generating System and various wind farms initially appeared promising but ultimately faced challenges with output consistency and transmission costs, failing to deliver expected benefits.

Large-scale projects have not lived up to their promises, burdened by inefficiencies and environmental harm. In contrast, rooftop solar offers a practical, community-based solution. By decentralizing energy production, rooftop solar minimizes land use and infrastructure needs and empowers individuals to contribute to a cleaner, resilient energy future. With its smaller environmental footprint and increased accessibility, rooftop solar is a more promising investment for America's energy future—one that bypasses the pitfalls of large-scale, corporate-driven projects.

Chapter 1

The Land Dilemma

America's pursuit of large-scale alternative energy projects has often come with a hidden cost—land. From sprawling solar farms to expansive wind farms, these projects require vast tracts of land, transforming open landscapes and sensitive ecosystems into industrial zones. This is the land dilemma: the quest for clean energy collides with the need to preserve natural environments. While renewable energy is vital to reducing our reliance on fossil fuels, the massive scale of these projects brings significant environmental trade-offs. What gets lost in the push for mega-projects is the fact that we are sacrificing one of our most precious resources—land—in the name of progress. This chapter explores the unintended consequences of America's land-hungry energy strategy, and why decentralized solutions like rooftop solar offer a more sustainable path forward, one that harnesses the power of clean energy without compromising our landscapes or ecosystems.

The Vast Amounts of Land Consumed by Large Solar Farms and Wind Projects

Large-scale solar and wind farms require significant land to produce energy at the levels needed to replace traditional power plants. For example, a typical utility-scale solar farm can require up to 8 acres per megawatt (MW) of electricity, while wind farms often need even larger tracts of land due to the spacing required between turbines for optimal efficiency. The Ivanpah Solar Project in California, for example, spans 3,500 acres (NREL, 2020), and similarly, large wind farms in Texas occupy thousands of acres (American Wind Energy Association, 2021). While these projects generate substantial amounts of energy, the sheer land footprint needed for renewable energy development has raised concerns about sustainability and efficient land use.

Environmental and Economic Impacts of Land Use

The environmental impacts of large-scale renewable energy projects are complex and often extend beyond the immediate footprint of the installations. Solar farms, for instance, disrupt natural habitats, contributing to a significant loss of biodiversity as they replace diverse ecosystems with expansive, monoculture-like solar arrays. This transformation affects a wide range of species, from soil microbes to larger mammals, all of which play critical roles in local ecosystems. In addition, the intensive water usage in some solar projects, especially in arid regions, places further stress on water resources, impacting both local wildlife and agricultural viability.

Wind farms, while producing clean energy, can similarly disrupt ecosystems. Bird and bat populations are particularly vulnerable to turbine blades, leading to increased mortality rates, especially for species already endangered or threatened. These wind farms can also impact ground-dwelling species, which may experience habitat fragmentation or loss, as construction and maintenance roads alter the landscape. This disturbance is often felt acutely in rural or semi-natural areas, where human development was previously minimal.

Moreover, the vast land areas required for these projects often overlap with agricultural land or designated conservation areas. This overlap can create conflicts over land use and resource rights, pitting renewable energy needs against food production and biodiversity conservation. Farmers, in particular, may face economic strain as arable land is reduced, potentially affecting food supply and prices in the region. The opportunity cost of dedicating land to renewable energy—often at the expense of farming or other productive land uses—has sparked debates, especially in agricultural regions where land scarcity already poses challenges.

Economically, these projects can have complex effects on local communities. While construction brings temporary jobs, the long-term employment opportunities associated with renewable energy installations tend to be limited, providing fewer sustainable jobs than many residents might hope for, particularly in rural areas. This economic displacement, combined with the extensive land requirements, can lead to resentment among locals who feel they've sacrificed both land and livelihood for projects that offer little enduring benefit in return (Pasqualetti, 2019). The combined ecological and economic impacts illustrate the need for more thoughtful planning and community engagement in renewable energy development.

Case Studies of Disrupted Ecosystems and Communities Displaced by Large Projects

Throughout the United States, large-scale renewable energy projects have promised cleaner power, but their implementation has often come at a steep cost to local ecosystems and communities. Far from the gleaming vision of sustainability, these mega-projects have disrupted delicate habitats, displaced wildlife, and uprooted communities. From wind farms encroaching on critical migratory routes to vast solar fields transforming once-pristine deserts into industrial zones, the environmental and social toll of these initiatives has been profound. In this section, we will explore case studies of how large renewable energy projects—despite their green intentions—have caused irreversible damage to ecosystems and forced communities from their homes, raising serious questions about whether this is the right path for America's energy future.

Ivanpah Solar Electric Generating System (California): The Ivanpah Solar Electric Generating System spans 3,500 acres in the Mojave Desert and was designed to produce 392 MW of electricity, enough to power around 140,000 homes annually. However, the project has only delivered about 60% of its expected energy output due to several operational setbacks, including mechanical issues and an over-reliance on natural gas to supplement solar power during periods of low sunlight. Environmentally, the project has faced criticism for its impact on local wildlife, particularly the desert tortoise, an endangered species. The construction disturbed tortoise habitats, forcing relocation efforts that were costly and not always successful. Additionally, the broader ecosystem was negatively affected, with concerns over the long-term damage to the fragile desert environment.

Data	Details
Total Land Use	3,500 acres
Promised Output	392 MW
Actual Output	60% of target
Land Impact	Desert tortoise displacement, ecosystem disruption

The project has also been criticized for the unintended consequences of its solar power technology. The intense heat generated by the mirrors has been reported to cause bird fatalities, as birds flying through the solar flux are severely burned. Despite its intentions to produce clean energy, these environmental impacts cast doubt on the long-term sustainability and ecological footprint of such large-scale solar projects, especially in sensitive ecosystems like the Mojave Desert.

Roscoe Wind Farm (Texas): The Roscoe Wind Farm, spanning 100,000 acres, stands as one of the largest wind farms globally, delivering 781.5 MW of power to about 250,000 homes. While it has succeeded in meeting its energy production targets, the land requirements for the turbines have significantly reduced available agricultural land, leading to conflicts with local farmers who rely on that land for cultivation and grazing. Furthermore, the wind turbines have had detrimental effects on wildlife, particularly birds and bats, many of which are killed or displaced by the turbines' rotating blades.

Data	Details
Total Land Use	100,000 acres
Promised Output	781.5 MW
Actual Output	Largely met goals
Land Impact	Agricultural land loss, wildlife disruption

Beyond its impact on farming and wildlife, the Roscoe Wind Farm also highlights a broader issue in large-scale renewable projects: balancing energy production with land-use efficiency. The vast expanse of land needed for wind farms like Roscoe raises questions about the long-term sustainability of such projects, particularly in rural areas where farming is a critical economic activity. Although renewable energy is a key solution to reducing carbon emissions, the trade-offs in land use and biodiversity continue to generate debate within the environmental and agricultural communities.

Crescent Dunes Solar Energy Project (Nevada): The **Crescent Dunes Solar Energy Project**, a high-profile solar thermal plant, was built on 1,670 acres in Nevada with the goal of generating 110 MW of solar power for 75,000 homes. However, the project faced persistent technical failures, particularly with its molten salt energy storage system, which severely limited its operational capacity. After years of underperformance and failing to deliver on its promises, the plant was officially shut down in 2020. This left behind concerns about the environmental impact on the desert ecosystem and the absence of a clear land restoration plan post-decommissioning.

Data	Details
Total Land Use	1,670 acres
Promised Output	110 MW
Actual Output	Never reached full capacity
Land Impact	Ecosystem disruption, lack of restoration plans

The land disturbance caused by Crescent Dunes includes not only the alteration of native landscapes but also potential long-term damage due to the intensive construction required for the facility. The lack of a comprehensive post-decommissioning strategy has raised alarms about whether the land will be rehabilitated or left as a blighted site in the desert. Additionally, the significant taxpayer investment in the project—which reached over $700 million in federal loan guarantees—underscores the economic risks involved in such large-scale projects, particularly when they fail to meet their expected outcomes (DOE, 2020).

This case is often cited as an example of the financial and ecological risks tied to large renewable energy projects, where the promise of clean energy is undermined by technological and operational limitations. It also illustrates the broader challenges of scaling up solar energy production without creating lasting negative impacts on the environment and public funds.

These case studies highlight the significant ecological costs and economic challenges associated with large-scale renewable energy projects. Although these projects produce clean energy, their vast land use results in the displacement of local ecosystems, which can include endangered species, and often disrupts agricultural land and rural communities. The long-term sustainability of such projects is questioned when they fail to meet energy production targets, as seen in the cases of Ivanpah and Crescent Dunes. Additionally, the environmental impacts, such as habitat destruction and the complexity of land restoration, make these projects less attractive compared to decentralized alternatives like rooftop solar.

The economic challenges are also pronounced. Local communities may face reduced agricultural productivity, diminished land value, or increased conflict over land rights due to large projects. The reliance on taxpayer funding for these projects, combined with their underperformance and higher-than-expected maintenance costs, calls into question the economic feasibility of large renewable farms. This makes decentralized, localized solutions like rooftop solar more appealing as they require less land, produce energy closer to where it is consumed, and avoid many of the negative ecological and economic impacts of large-scale projects.

Chapter 2

The High Costs of Scale

As America rushes to embrace large-scale renewable energy projects, the staggering costs of these ventures are often overlooked or downplayed. In this Chapter, "The High Costs of Scale," we will dive into the financial, environmental, and societal expenses tied to these massive initiatives. While these projects are marketed as solutions to the country's energy crisis, their reality is far more complex. From ballooning budgets and extended timelines to hidden environmental degradation and the displacement of local communities, the true costs of scaling up renewable energy are steep. This chapter explores how the pursuit of size in renewable energy has led to inefficiencies, waste, and unintended consequences, making the case for a more localized, cost-effective approach like rooftop solar.

The Hidden Costs of Building and Maintaining Large Alternative Energy Projects

While large-scale alternative energy projects such as solar farms and wind turbines are often lauded for their potential to generate significant amounts of renewable energy, they come with a plethora of hidden costs that undermine their economic and environmental benefits. These hidden costs extend beyond the initial capital expenditure and include long-term maintenance, environmental mitigation, and unforeseen operational expenses.

One of the primary hidden costs is the environmental mitigation required to comply with regulatory standards. Large projects often necessitate extensive environmental impact assessments and the implementation of measures to protect local wildlife and ecosystems. For instance, the Ivanpah Solar Electric Generating System in California required substantial efforts to relocate endangered desert tortoises and restore disrupted habitats (Lovich & Ennen, 2017). These mitigation efforts not only increase the overall cost but also delay project timelines, affecting the return on investment.

Maintenance costs for large-scale projects are another significant factor. Wind turbines and solar panels require regular upkeep to ensure optimal performance. Wind farms, for example, face mechanical issues such as gearbox failures and blade erosion, which necessitate costly repairs and replacements (Wang & Wang, 2020). Similarly, solar farms like Crescent Dunes have experienced persistent operational failures, leading to increased maintenance expenses and eventual project shutdowns (U.S. Department of Energy, 2020). These ongoing costs can erode the financial viability of large projects, making them less attractive compared to smaller, decentralized alternatives.

Additionally, the integration of large-scale projects into existing energy systems often reveals unforeseen operational costs. These include the need for backup power sources to manage the intermittency of renewable energy generation. The Ivanpah project, for instance, had to rely on natural gas to supplement solar energy during periods of low sunlight, increasing operational costs and diminishing the environmental benefits (NREL, 2020). Such dependencies highlight the complexities and additional expenses associated with maintaining consistent energy output from large-scale renewable projects.

Infrastructure and Transmission Challenges

Infrastructure and transmission challenges pose significant barriers to the successful implementation and operation of large-scale alternative energy projects (see table below). The geographic location of many renewable energy sources often necessitates the construction of extensive transmission lines to deliver electricity to urban centers, which incurs substantial costs and logistical complexities.

Large solar and wind farms are frequently situated in remote areas where land is abundant and wind or sunlight is optimal. However, these locations are typically far from major population centers, requiring the development of new transmission infrastructure to transport the generated electricity. Building these transmission lines involves high capital investments and can face substantial delays due to regulatory approvals, environmental concerns, and local opposition (Cohen, 2019). For example, the Roscoe Wind Farm in Texas required significant investment in transmission infrastructure to connect to the grid, adding to the overall cost and complexity of the project (American Wind Energy Association, 2021).

Infrastructure Challenge Breakdown	Details
Transmission Distance	Remote generation sites far from urban centers
Energy Loss in Transmission	Decreased efficiency due to energy loss over long distances
Logistical/Regulatory Hurdles	Negotiations, environmental regulations, local resistance

Moreover, the existing grid infrastructure in many regions is not equipped to handle the variable and decentralized nature of renewable energy sources. Integrating large amounts of wind and solar power can strain the grid, leading to inefficiencies and increased likelihood of outages. The Texas energy grid, during the 2021 winter storm, demonstrated the vulnerabilities of relying heavily on renewable sources without adequate grid support and backup systems (Wang & Wang, 2020). These infrastructure challenges necessitate additional investments in grid modernization and energy storage solutions, further escalating the costs associated with large-scale renewable energy projects.

Transmission losses are another critical issue. Electricity loses energy as it travels long distances through power lines, reducing the overall efficiency of large-scale projects. These losses can be significant, particularly for projects located far from end-users, thereby diminishing the net energy output and economic benefits of the projects (U.S. Department of Energy, 2019). Addressing these transmission challenges requires innovative solutions such as high-voltage direct current (HVDC) lines and localized energy storage, which, while effective, add to the financial burden of renewable energy projects.

Financial Inefficiencies in Large-Scale Energy Solutions

Financial inefficiencies inherent in large-scale alternative energy projects further undermine their viability and cost-effectiveness. These inefficiencies stem from factors such as high initial capital costs, long payback periods, and vulnerability to policy changes and market fluctuations.

The initial capital investment required for large renewable energy projects is substantial. Building extensive solar farms or wind turbines involves significant upfront costs for land acquisition, equipment, installation, and infrastructure development. For instance, the Crescent Dunes Solar Energy Project in Nevada incurred over $700 million in federal loan guarantees, reflecting the massive financial commitment required (U.S. Department of Energy, 2020). These high initial costs create barriers to entry and make large projects financially risky, especially when projected energy outputs are not met, as seen with Crescent Dunes.

Long payback periods are another financial inefficiency associated with large-scale projects. The time required to recoup the initial investment through energy sales and incentives can span decades, making these projects less attractive compared to smaller, more rapidly deployable solutions. The delayed return on investment is exacerbated by operational inefficiencies and maintenance costs, as previously discussed, further prolonging the payback period and increasing financial risk (Pasqualetti, 2019).

Moreover, large-scale renewable energy projects are highly sensitive to policy changes and market conditions. Subsidies, tax incentives, and regulatory support are often crucial for the financial viability of these projects. However, shifts in government policies or economic downturns can significantly impact project funding and profitability. The reliance on government support makes large projects vulnerable to political instability and changes in public policy, adding an additional layer of financial uncertainty (Sovacool, 2017).

Another often overlooked financial challenge is the hidden costs associated with grid integration and infrastructure upgrades necessary to handle energy from large renewable installations. Many of these projects are located in remote areas, where electricity demand is low but land availability is high. This geographic separation requires substantial investments in transmission lines and grid management technologies to transport power to population centers, significantly adding to overall project expenses. Furthermore, fluctuating energy output from renewable sources like solar and wind necessitates the construction of backup power plants or storage systems to ensure grid stability, further compounding costs. Together, these hidden expenses highlight the financial complexities of large-scale renewable projects and underscore the importance of considering a balanced, diversified approach to energy generation.

Financial Inefficiency Breakdown	**Details**
Dependency on Subsidies	Projects rely on government subsidies and incentives to be financially viable
Intermittency Issues	Need for fossil-fuel-based backup systems to address fluctuating energy output
Subsidies Mask True Costs	Financial support hides real construction and operational expenses

In contrast, decentralized solutions like rooftop solar offer greater financial flexibility and resilience. Rooftop installations typically require lower initial investments, shorter payback periods, and reduced dependence on extensive infrastructure and subsidies. Homeowners and businesses can benefit directly from energy savings and incentives, making rooftop solar a more economically efficient and less risky investment compared to large-scale projects (Baker, 2022).

The high costs associated with large-scale alternative energy projects present significant challenges that undermine their economic and environmental benefits. Hidden costs related to environmental mitigation, maintenance, and operational dependencies, coupled with substantial infrastructure and transmission challenges, contribute to financial inefficiencies that make these projects less viable in the long term. These issues highlight the need for more sustainable and economically efficient energy solutions, such as rooftop solar, which offer localized benefits without the extensive land use and financial burdens inherent in large-scale renewable energy projects.

Chapter 3

Alternative Energy's Biggest Failures Lessons and Taxpayer Burdens

In the quest for a sustainable future, America has invested billions of taxpayer dollars into alternative energy projects, with the hope of reducing reliance on fossil fuels, cutting greenhouse gas emissions, and bolstering energy independence. These investments aimed to usher in a cleaner, more resilient energy landscape, transforming the nation's power grid to meet modern environmental and economic demands. While many projects have succeeded and demonstrated the potential of renewable technologies, a notable number have not lived up to their promises. Instead, these projects have highlighted the substantial risks involved in large scale renewable energy development, where technical challenges, high costs, and environmental impacts often complicate lofty ambitions.

The projects examined in this chapter were once symbols of renewable energy innovation, positioned as breakthroughs that would drive significant change. Yet, their failures underscored serious miscalculations, from underestimating operational and maintenance costs to overlooking the complexities of scaling unproven technologies. Consequently, taxpayers were left footing the bill for these projects, as investments failed to yield the promised returns in clean energy. Beyond the financial losses, these high-profile missteps have had a broader impact on public trust. Each failed project has not only drained public resources but has also cast a shadow over the renewable energy sector, fueling skepticism among policymakers and citizens alike.

These setbacks have tarnished the reputation of renewable energy efforts, with critics using them as examples of why large-scale renewable energy investments are unreliable or wasteful. As a result, the renewable energy sector faces increased scrutiny, with failed projects becoming a point of contention in the debate over future energy investments. This chapter examines these projects, exploring the causes behind their failures and the lessons they offer. It reveals the importance of careful planning, realistic projections, and strong oversight in public investments, as these are essential to ensuring that renewable energy projects not only meet their environmental goals but also remain financially responsible and viable in the long term

The Ivanpah Solar Electric Generating System – A $2.2 Billion Shortfall

Ivanpah, a concentrated solar power (CSP) project launched with great fanfare in California's Mojave Desert, stands as one of the most infamous alternative energy failures. Designed with a capacity of 392 megawatts (MW), Ivanpah was expected to generate enough electricity to power over 100,000 homes. However, the facility has consistently fallen short of these goals, struggling to produce energy at its promised levels. The challenges it faces stem in part from its dependence on natural gas to start up daily, a requirement that significantly undermines its intended environmental benefits, as burning natural gas adds to the facility's carbon footprint (BrightSource Energy, 2021).

Initially backed by $1.6 billion in federal loans and taxpayer subsidies, Ivanpah has come under scrutiny for its ongoing struggles to meet performance expectations, prompting questions about the viability of large-scale CSP projects as a whole. Its extensive 3,500-acre footprint in the Mojave Desert has also attracted environmental concerns. The project disrupted sensitive desert ecosystems, impacting species such as the threatened desert tortoise and introducing a risk to bird populations, which can suffer fatal burns upon flying into concentrated beams of sunlight. These issues have compounded the project's financial and operational struggles, tarnishing Ivanpah's reputation and raising broader concerns about the environmental costs and sustainability of large-scale solar farms (California Energy Commission, 2023).

Solyndra – The $535 Million Collapse

Perhaps no alternative energy project has had a more spectacular fall from grace than Solyndra, a California-based solar panel manufacturer that once promised to revolutionize the solar industry. Armed with an innovative but costly cylindrical solar panel design, Solyndra aimed to dominate the market, positioning itself as a trailblazer in renewable technology. Backed by $535 million in federal loans through the Department of Energy, Solyndra was expected to lead the charge in sustainable energy solutions. However, just two years after receiving this substantial funding, Solyndra filed for bankruptcy, leaving thousands of panels unsold, factories idle, and hundreds of workers without jobs (U.S. Department of Energy, 2012).

The collapse was a severe blow to the renewable energy sector, igniting public criticism and triggering extensive investigations into the Department of Energy's loan process. The Solyndra debacle highlighted the volatility and risks associated with taxpayer-funded investments in unproven technologies, especially in a competitive and rapidly evolving market where innovation alone is not enough. This high-profile failure became a cautionary tale in renewable energy funding, underscoring the critical need for rigorous oversight, realistic market assessments, and careful planning when allocating public funds. Today, Solyndra serves as a reminder of the importance of due diligence and grounded projections to ensure that federal investments genuinely advance renewable energy goals without imposing undue risk on taxpayers.

Crescent Dunes Solar Energy Project – A $1 Billion Miscalculation

The Crescent Dunes project in Nevada was another ambitious CSP (concentrated solar power) facility designed to store solar energy in molten salt, allowing it to generate 110 MW of power, even after sunset—a capability that would have represented a breakthrough in renewable energy storage. With its potential to overcome the intermittency issues that often limit solar energy, Crescent Dunes quickly became a beacon of hope for reliable, large-scale renewable energy storage. Backed by $737 million in federal loans, Crescent Dunes embodied high expectations and significant taxpayer investment. However, the project soon encountered a series of setbacks: technical difficulties, frequent shutdowns, and power output that fell well below projections became persistent issues, leading to operational failures that impacted the plant's financial viability and its capacity to deliver on promises (U.S. Department of Energy, 2020).

One of the project's most significant setbacks was its inability to fulfill contractual obligations to its primary customer, NV Energy, resulting in a breach of contract that further hindered its operations. Despite the bold promise of commercial success, Crescent Dunes failed to achieve the reliability and scalability necessary to make it a viable energy source, ultimately leaving the plant effectively defunct by 2019. The project's collapse marked a substantial loss for taxpayers and investors alike, raising doubts about the feasibility of large-scale CSP and thermal storage solutions. This high-profile failure has since cast a shadow over future CSP initiatives, emphasizing the importance of realistic technological assessments and robust contingency planning in renewable energy projects.

Cape Wind Project – An Offshore Wind Fiasco

The Cape Wind Project off the coast of Cape Cod was intended to be America's first offshore wind farm, designed to produce 468 MW of renewable energy and serve as a pioneering model for offshore wind in the U.S. Supported by $150 million in tax credits and federal support, Cape Wind seemed poised to become a landmark project in renewable energy. However, the project quickly became entangled in fierce opposition from local communities, environmental organizations, and influential political figures, leading to protracted litigation, regulatory hurdles, and costly delays (Cape Wind Associates, 2021).

Over a decade of stalled development drained resources and pushed costs ever higher, ultimately resulting in the project's official cancellation in 2017. Not only did Cape Wind fail to deliver clean energy to American households, but it also set a difficult precedent for future offshore wind projects, highlighting the complex political and logistical challenges that such projects face in the U.S. The cancellation left taxpayers with nothing to show for their investment aside from legal expenses and a legacy of lost potential. Cape Wind underscored the importance of securing strong community and political backing for renewable projects, as well as the need for streamlined regulatory processes to prevent similar costly failures in the future

Abound Solar – A $70 Million Collapse in Manufacturing

Abound Solar, a Colorado-based manufacturer specializing in thin-film solar panels, received $70 million in federal loans with high expectations of producing affordable, high-efficiency solar technology. The company aimed to make solar energy more accessible, yet it quickly encountered major quality control issues. Reports surfaced of defective panels that degraded rapidly, falling far short of promised performance standards and rendering the panels effectively unusable (U.S. House of Representatives Committee on Energy and Commerce, 2012).

By 2012, Abound Solar declared bankruptcy, leaving behind a vast inventory of faulty panels and significant environmental cleanup costs at its abandoned facilities. This high-profile failure underscored the risks of federal financing for unproven technologies, especially when production quality and durability standards are not assured. The financial and environmental aftermath of Abound Solar's collapse prompted serious scrutiny of taxpayer funding for similar ventures, highlighting the need for rigorous assessments of technological viability and oversight in renewable energy investments. This case served as a cautionary example of the challenges involved in supporting early-stage technologies with public funds and the potential costs when such projects go awry

Beacon Power – The $43 Million Flywheel Disaster

Beacon Power, a Massachusetts-based company, aimed to enhance grid stability through innovative flywheel energy storage technology, a system designed to store and release excess energy during peak demand hours. With $43 million in federal funding, Beacon set out to construct a 20 MW storage plant in New York, anticipating that this facility would reduce grid strain and serve as a model for advanced energy storage. However, despite the promise of its technology, Beacon quickly encountered severe financial challenges, and less than a year after receiving federal support, the company declared bankruptcy (Federal Energy Regulatory Commission [FERC], 2013).

Beacon's failure highlighted both the financial and technical challenges inherent in energy storage solutions, a sector where costs and scalability remain significant obstacles. The collapse raised concerns about investing taxpayer funds in experimental technologies that lack established commercial viability. As a result, Beacon became a case study illustrating the need for cautious investment in unproven innovations and underscored the importance of thorough vetting and risk assessment in public funding for energy projects. The project's quick demise served as a reminder of the financial risks associated with energy storage ventures and the need for a clear path to profitability and sustainability before committing substantial taxpayer resources

Lessons Learned from Alternative Energy Failures

Each of these projects exemplifies the risks and challenges inherent in large-scale renewable energy ventures funded by taxpayer dollars. While these initiatives were driven by a genuine need to advance clean energy and reduce dependence on fossil fuels, their failures underscore critical issues within the sector: an over-reliance on untested technologies, insufficient oversight, and a tendency to overestimate the immediate impact of ambitious alternative energy projects. The billions of taxpayer dollars funneled into these initiatives serve as stark reminders that the journey toward a sustainable energy future requires caution, accountability, and realistic expectations.

As the renewable energy landscape continues to evolve, it's essential for policymakers, companies, and investors to reflect on these costly missteps to avoid repeating them. The rush to support large-scale projects has often overshadowed the need to build a stable foundation with proven, scalable, and economically viable technologies. By emphasizing alternatives like rooftop solar, energy-efficient upgrades, and community-centered renewable solutions, we could develop a more resilient and sustainable path forward. These options allow for incremental growth that aligns with existing infrastructure and supports local economies, reducing the risk of costly oversights and fostering more gradual, reliable progress.

Moreover, the high-profile failures of these projects should be leveraged as valuable case studies, guiding future investments toward responsible, community-driven, and scientifically validated renewable energy strategies. As renewable technology continues to advance, a critical balance must be struck between innovation and feasibility. Policymakers and investors need to prioritize projects with realistic timelines, measurable outcomes, and transparent oversight mechanisms that ensure public funds are used effectively.

Ultimately, by embracing a diversified approach to energy development, including smaller-scale and distributed generation solutions, we can cultivate a renewable energy landscape that is less vulnerable to single points of failure. This approach promotes energy independence and resilience at the community level, helping to ensure that the promise of a sustainable future becomes a reality—grounded not in speculation, but in scientifically sound, economically sustainable practices that can withstand the challenges of a rapidly changing world.

Chapter 4

Efficiency vs. Scale

In the race to solve America's energy challenges, bigger isn't always better. In this Chapter, "Efficiency vs. Scale," we will examine the critical trade-offs between the massive scale of large renewable energy projects and the efficiency of decentralized solutions like rooftop solar. While large-scale wind farms and solar fields promise high energy output, they are often hampered by transmission losses, environmental disruption, and escalating costs. In contrast, smaller, decentralized systems such as rooftop solar deliver energy precisely where it's needed, reducing inefficiencies and offering a more sustainable, localized solution. This chapter explores why, in the quest for clean energy, efficiency often trumps scale, and how embracing smaller, distributed systems could be the key to a more effective and equitable energy future.

Why Bigger Isn't Always Better: Diminishing Returns of Large Projects

While large-scale renewable energy projects such as wind farms and solar fields promise substantial energy output, they often encounter diminishing returns as they expand. The initial allure of these projects lies in the potential for high energy generation across vast spaces; however, the costs of building and maintaining such extensive infrastructure often grow disproportionately compared to the energy produced. This is partly because large-scale projects demand enormous amounts of land, materials, and a sizable workforce. Unfortunately, the energy output per unit of land can decrease as more land is developed, primarily due to resource constraints and declining site quality (Tomaschek & Miller, 2021). For example, the first wind turbines in a wind farm can often be positioned in optimal wind corridors, but as expansion continues, less ideal locations are inevitably utilized, resulting in lower overall efficiency and output. This phenomenon is evident in both wind and solar projects, where increasing scale does not always equate to proportionally higher energy production.

Additionally, larger projects bring more complex maintenance requirements. The sheer distances between turbines or solar arrays introduce logistical challenges, making repairs and routine upkeep more demanding and costly. Transportation for maintenance crews and equipment over large areas adds to operational expenses and lengthens response times for repairs, ultimately increasing downtime and reducing energy production consistency. These issues collectively result in a situation where the marginal cost of each additional unit of energy rises, further eroding the financial benefits of scaling up (Joskow, 2019). Compounding this, larger projects are often subject to regulatory and environmental review processes, which can delay construction and escalate costs, particularly when projects encroach on protected lands or conflict with local land use priorities.

In contrast, smaller, decentralized systems like rooftop solar installations can be scaled with greater efficiency, avoiding many of the logistical pitfalls associated with sprawling projects. Rooftop solar, for example, leverages existing infrastructure, reducing the need for additional land or extensive new transmission systems. Such installations also allow energy generation closer to where it will be used, minimizing transmission losses and infrastructure requirements. Decentralized systems are typically less susceptible to land-use conflicts and can integrate more seamlessly into local energy grids. This approach not only enhances efficiency but also distributes energy production more equitably across regions, empowering communities to participate directly in the transition to renewable energy. As a result, smaller, community-centered renewable solutions may offer a more viable and resilient path forward, capitalizing on the strengths of renewable technology without the drawbacks of excessive scale.

Comparing Energy Output per Square Foot: Large Projects vs. Rooftop Solar

One of the key metrics that reveals the inefficiency of large-scale renewable projects is energy output per square foot of land. Large solar farms, for instance, require expansive areas to achieve the same energy production that could be achieved through numerous rooftop solar installations spread across urban environments. The National Renewable Energy Laboratory (NREL) reports that utility-scale solar projects typically generate around 10-12 watts per square foot, while rooftop solar can produce around 15-20 watts per square foot, depending on location and orientation (NREL, 2020). This disparity is largely due to the fact that rooftop solar takes advantage of already existing infrastructure—roofs—thereby eliminating the need to clear additional land or build extensive new infrastructure.

Diminishing Returns in Large Projects	Details
Increased Land Use	Larger projects require more land, often in suboptimal locations
Higher Maintenance Costs	Greater distances between infrastructure make repairs more costly
Marginal Cost Increases	Costs rise disproportionately as scale increases

Rooftop solar also circumvents the significant environmental challenges that accompany large-scale renewable projects, such as habitat fragmentation, biodiversity loss, and disruption of local ecosystems. By using already-developed spaces like rooftops, it maximizes land efficiency without compromising surrounding habitats or interfering with valuable agricultural areas. Additionally, rooftop solar reduces the need for extensive transmission infrastructure, as the energy generated is consumed close to the source. This localized generation minimizes energy losses typically incurred during long-distance transmission and decreases the need for costly upgrades to the power grid, which are often required to accommodate remote energy sources. Furthermore, rooftop solar empowers homeowners and communities by allowing them to generate their own electricity, which fosters a sense of ownership and engagement in renewable energy initiatives. This self-sufficiency contributes to energy independence, making communities more resilient to fluctuations in the larger energy market and reducing reliance on centralized power systems. By integrating renewable energy into everyday life in a way that is accessible and practical, rooftop solar plays a vital role in democratizing energy, paving the way for a more sustainable and locally driven energy future.

Energy Output Comparison	Large Projects	Rooftop Solar
Energy Output (Watts per Square Foot)	10-12	15-20
Land Use	Requires vast open land	Uses existing rooftops
Environmental Impact	Disrupts ecosystems and agricultural land	Minimal additional impact

Energy Loss in Transmission from Remote Farms to Urban Centers

A significant downside of large-scale wind and solar farms is the distance between where energy is generated and where it is ultimately consumed. Many of the optimal locations for renewable energy projects—such as wind corridors in remote plains and sunny deserts—are far removed from major population centers where electricity demand is highest. Consequently, energy produced at these remote sites must be transmitted over long distances, introducing inefficiencies due to energy loss during transmission. According to the

U.S. Energy Information Administration (EIA), approximately 5-10% of the electricity generated at power plants is lost in transmission and distribution (EIA, 2021). In the case of renewable energy, this percentage may even rise when accounting for the challenges of transporting power from far-flung sites to urban centers.

These transmission losses become a substantial obstacle for large-scale renewable projects, as the energy must travel farther than conventional sources located closer to end users. The infrastructure needed to support this—namely high-voltage transmission lines, substations, and grid interconnections—further increases the cost, complexity, and environmental footprint of large-scale projects. These transmission lines can also have additional environmental impacts, including the need to clear land for right-of-way and the potential impact on wildlife corridors. Moreover, the process of securing land rights for transmission corridors can delay projects, create legal conflicts with local landowners, and face public opposition, adding to both time and expense.

In contrast, rooftop solar represents a localized energy solution, generating electricity near the point of consumption and thus minimizing transmission losses and infrastructure requirements (see table below). By producing energy on-site, rooftop solar eliminates the need for long-distance transmission lines and the energy losses associated with them. Additionally, it reduces dependence on large grid infrastructure, creating a more resilient energy system that is less vulnerable to grid disruptions. This proximity also empowers individual households and businesses to actively participate in clean energy generation, increasing local energy independence. Furthermore, rooftop solar can alleviate the strain on centralized grids, especially during peak demand periods, by providing supplementary power precisely where it's needed most. This efficiency and local impact make rooftop solar an appealing alternative, delivering renewable energy benefits directly to communities without the pitfalls of extended transmission needs.

Transmission Efficiency	Large Projects	Rooftop Solar
Energy Loss in Transmission	5-10%	Minimal
Distance to Urban Centers	Far from cities	On-site generation

While large-scale renewable energy projects, such as wind farms and solar fields, are often touted as key steps in transitioning away from fossil fuels, many of these ventures function as political pork-barrel projects. These projects are frequently backed by politicians seeking status wins and corporate gain, rather than genuine energy efficiency. Large corporations benefit from government subsidies and tax credits, while political leaders can claim environmental wins. However, these projects often fall short of their promises due to inefficiencies related to land use, maintenance, and energy transmission (Joskow, 2019).

For example, large wind and solar farms, which occupy vast areas of land, often underperform due to transmission losses that occur when energy is sent over long distances from remote areas to urban centers. Additionally, maintenance costs rise as these projects scale, requiring more complex and expensive infrastructure. The result is that the actual energy output per square foot is significantly lower than projected.

In contrast, rooftop solar is a far more efficient and cost-effective alternative. It provides energy where it's consumed, significantly reducing transmission losses and increasing overall efficiency. Rooftop solar installations also have minimal environmental impact because they utilize existing infrastructure (rooftops) rather than requiring vast new areas of land to be cleared. This decentralized energy model is cheaper to build, requires less maintenance, and does not rely on expensive transmission infrastructure.

Moreover, rooftop solar systems are inherently scalable. Homeowners and businesses can install these systems without needing to wait for massive government-backed projects to be approved and constructed. They can also benefit from financial incentives like net metering, which allows them to sell excess power back to the grid, increasing the economic return of their investment (NREL, 2020).

In contrast to the inefficiencies of large-scale projects, rooftop solar represents a practical and scalable energy solution that can be implemented by individuals and businesses alike. As such, it's clear that decentralized, rooftop solar solutions not only make more sense environmentally but also economically, without the corporate and political entanglements that often plague large-scale projects. Finally, while large-scale renewable energy projects are often framed as vital steps toward sustainability, they are frequently inefficient and driven by political and corporate agendas rather than genuine environmental progress. In contrast, rooftop solar provides a more sustainable, efficient, and cost-effective solution with fewer environmental impacts and greater potential for widespread adoption.

Chapter 5

The Desert Myth
Large Solar Farms in Deserts Are Not Harmless as They Seem

The myth that large solar farms in deserts are harmless is a widely held belief that vast desert landscapes are perfect, impact-free locations for large-scale solar farms. While deserts may seem barren and lifeless, ideal for renewable energy projects, the truth is far more nuanced. These fragile ecosystems, often overlooked, are home to unique wildlife, vegetation, and delicate ecological balances. Large solar installations not only disrupt these environments but also demand vast amounts of land, water for cleaning, and extensive infrastructure, causing irreversible changes to the landscape. This chapter reveals the hidden environmental costs of desert solar farms, debunking the myth of their harmlessness, and highlights why decentralized options like rooftop solar offer a more sustainable solution.

Environmental Degradation in Desert Ecosystems

Deserts are often perceived as barren, uninhabited landscapes, seemingly ideal for large-scale solar projects due to their abundant sunlight and open spaces. However, this perception overlooks the complex and fragile ecosystems that deserts support. These arid environments are home to a diverse range of plants, animals, and microorganisms uniquely adapted to extreme conditions, often thriving in delicate balances that can be easily disrupted. The construction of solar farms in deserts can lead to significant soil compaction, which affects the soil's ability to retain water—a critical function in arid environments where moisture is scarce. Compacted soils lose their natural porosity, making it difficult for rainwater to seep in and recharge subsurface moisture levels, a process that can take decades to recover (Lovich & Ennen, 2017).

Moreover, the disruption caused by solar installations extends to natural drainage patterns, as these large arrays often alter the landscape by creating obstacles to the natural flow of water. This redirection of water can lead to increased erosion in some areas and reduced water availability in others, severely impacting native vegetation that relies on minimal but crucial seasonal rainfall. For instance, desert plants, such as the Joshua tree in the Mojave Desert, have evolved to survive in specific microclimates and are sensitive to changes in soil composition and moisture levels. Once disturbed, these ecosystems may struggle to recover due to the slow growth rates of desert flora and the long-lived nature of many desert species.

The Mojave Desert is a prime example of a desert ecosystem hosting a variety of unique and often endangered species, such as the desert tortoise, Mojave ground squirrel, and several specialized lizard species. These animals depend on the sparse vegetation for food, shelter, and temperature regulation. When large tracts of land are converted into solar farms, their habitats are fragmented or eliminated, leaving them vulnerable to predation and environmental stressors. The industrialization of desert lands not only threatens these native species but also disrupts the roles they play in the broader ecosystem. For example, desert tortoises contribute to soil health through their burrowing behavior, creating microhabitats that support smaller organisms.

The cumulative effect of these disruptions is the degradation of entire desert ecosystems, which are already vulnerable due to their limited water resources and harsh climate. As solar projects expand, these ecosystems face increasing pressure, challenging the sustainability of placing renewable energy installations in such sensitive regions. Thoughtful planning and environmental assessments are critical to balancing the need for renewable energy with the preservation of desert ecosystems, as failing to do so may result in irreversible impacts on these biodiverse and resilient landscapes.

Disrupting Fragile Habitats and Native Species

Solar farms in deserts pose a serious threat to fragile habitats and native species, transforming once-undisturbed landscapes into industrial zones that disrupt ecological balance. Large projects such as the Ivanpah Solar Facility in the Mojave Desert, for example, have led to significant habitat disruption for the endangered desert tortoise. Efforts to relocate these tortoises have been costly, and studies indicate that relocation often yields limited success due to the animals' territorial nature and complex habitat needs, leading to increased mortality rates and disorientation (U.S. Department of Energy, 2020).

Beyond tortoises, many other species, including birds, insects, and small mammals, depend on the sparse but essential vegetation and open spaces that characterize desert ecosystems. Birds, particularly, are at risk, as solar arrays and mirrors can create reflective surfaces that resemble bodies of water, confusing and attracting them—a phenomenon known as the "lake effect"—resulting in fatal collisions or exposure to extreme heat. Insects, too, play vital roles as pollinators, but the alteration of their habitats can reduce plant reproduction and ultimately affect the food web that sustains other species.

The introduction of large-scale solar infrastructure also disrupts natural soil structure, compacts the ground, and alters drainage patterns, fundamentally changing the conditions that native plants and animals rely on to survive. Desert plants are often long-lived, slow-growing, and highly adapted to their environment, making them particularly vulnerable to disturbances. Once vegetation is removed or disturbed for development, erosion becomes a severe problem, and soil recovery can take decades, further stressing remaining wildlife and reducing biodiversity. This loss of vegetation and habitat fragmentation creates barriers for animals that need to roam widely to find food and mates, putting additional pressure on already vulnerable populations.

In the long term, these changes can lead to cascading ecological impacts, where the loss of one species affects others, gradually eroding the biodiversity that makes desert ecosystems unique and resilient. The potential for long-term ecological damage is considerable, especially given that desert ecosystems are slow to recover from disturbances. Without careful planning, large solar projects may bring about a decline in desert biodiversity, challenging the sustainability of using these sensitive landscapes for renewable energy. Balancing the urgent need for clean energy with the equally critical need to protect fragile desert ecosystems requires a nuanced approach that minimizes habitat disruption and considers alternative, less ecologically intrusive locations for solar infrastructure.

Farmlands at Risk: How Wind and Solar Projects Encroach on Agricultural Land

As the push for renewable energy accelerates, a growing concern is emerging: the encroachment of wind and solar projects on valuable agricultural land. In recent years, farmland—once reserved for food production—has increasingly been repurposed to host large-scale renewable energy developments, leading to a significant debate over land use. "Farmlands at Risk: How Wind and Solar Projects Encroach on Agricultural Land" examines this emerging conflict, illustrating how clean energy initiatives, though essential for reducing greenhouse gas emissions, may inadvertently compromise food security and rural livelihoods.

The shift toward placing renewable projects on fertile farmland brings with it a challenging trade-off. As wind turbines and solar panels consume vast swathes of agricultural land, farmers face declining access to viable plots, impacting crop yields and potentially driving up food prices. For rural communities that rely heavily on farming, the economic impact is considerable; land previously devoted to food production now contributes to renewable energy, potentially altering local economies and even leading to depopulation in farming regions. In areas where the land is both fertile and well-suited for energy projects, farmers and policymakers alike grapple with difficult decisions about how to allocate limited resources without undermining one sector for the benefit of another.

This encroachment is particularly concerning in regions facing heightened demand for both food and energy, as population growth continues and climate change disrupts agricultural patterns. Renewable energy installations often create a permanent or semi-permanent footprint, making it challenging to restore the land for agriculture after decommissioning. Soil compaction, changes to drainage, and altered microclimates due to the shadow effects of turbines or solar panels can also affect the quality of the land for future farming. The risks of sacrificing productive agricultural spaces for energy infrastructure are therefore not only immediate but potentially enduring, with implications for food production that could last well beyond the lifespan of any single energy project.

Ultimately, this section underscores the importance of a balanced approach to land use—one that prioritizes food security and supports rural economies while still advancing renewable energy goals. Solutions may include directing renewable projects toward less productive land or integrating dual-use systems where agriculture and energy generation coexist. Policymakers and industry leaders must work collaboratively to develop strategies that protect prime farmland from overdevelopment, ensuring that the pursuit of clean energy does not come at the cost of future food security.

Loss of Productive Farmland to Large-Scale Projects

Wind and solar energy projects are increasingly being developed not only in deserts but also on farmlands, leading to the loss of productive agricultural land. In regions like the Texas Panhandle, large wind farms such as the Roscoe Wind Farm have consumed vast tracts of land previously dedicated to crop cultivation and livestock grazing (Carolan, 2018). While leasing land for wind turbines provides additional income for some farmers, others face restricted access to essential areas needed for food production. This shift in land use places pressure on agricultural output, reducing the amount of farmland available for crops and potentially lowering overall food production.

As a result, the expansion of renewable energy projects into agricultural areas introduces long-term economic consequences for rural communities. Reduced access to fertile land impacts not only crop yields but also the local food supply chain, driving up food costs and affecting the viability of rural economies dependent on agriculture. The gradual displacement of farming activities by energy infrastructure can disrupt local markets and alter the cultural fabric of farming communities, as fewer families are able to sustain traditional agricultural practices. In areas where land is in high demand, particularly in fertile regions, the push to prioritize renewable energy may inadvertently harm food security and reduce rural economic resilience.

Moreover, the impact of wind and solar installations on farmland extends beyond immediate land loss. Construction activities, including soil compaction, drainage alteration, and permanent infrastructure, can degrade soil health and hinder future agricultural productivity. As these renewable energy projects are often long-term investments, the possibility of restoring the land to its former agricultural use becomes increasingly challenging. Policymakers and industry stakeholders face a complex dilemma: how to balance the urgent need for clean energy with the equally critical need to protect agricultural resources. A more strategic approach to land allocation, one that includes siting projects on less arable or marginal lands, may be essential to ensure that the expansion of renewable energy does not come at the cost of food production and rural stability.

Economic Impacts on Local Farmers and Rural Communities

The economic consequences for farmers and rural communities can be substantial, often presenting a double-edged sword. On one hand, some landowners benefit financially by leasing portions of their land to energy companies, generating a steady income stream that can help offset fluctuating crop revenues. However, for many farmers, the land taken up by renewable energy projects translates into a direct reduction in productive acreage, ultimately lowering crop yields and limiting agricultural output. This decrease in available farmland can strain farmers' profitability, forcing them to make difficult choices about how to allocate resources, and reducing their capacity to compete in an already challenging agricultural market (Pasqualetti, 2019).

In the long run, the presence of large-scale energy infrastructure may also impact property values, as the landscape changes from agricultural to industrial. This shift can make it difficult for farmers to sell or expand their operations, affecting generational farming continuity and reducing the overall attractiveness of the land for future agricultural use. For rural communities, the economic repercussions extend beyond individual farms; as farming income declines, so too does the vitality of local businesses and services that depend on a thriving agricultural sector. The influx of renewable energy projects, while beneficial to energy goals, rarely brings a corresponding increase in local job opportunities. Wind and solar farms typically require significant labor during the construction phase, but once operational, they create few permanent jobs. As a result, the expected long-term economic benefits for rural areas are minimal, leaving communities to bear the environmental and economic costs without substantial financial gain.

Over time, this economic imbalance can lead to a decline in rural economies, as families find it increasingly difficult to sustain livelihoods tied to farming. The lack of permanent job creation coupled with reduced agricultural productivity can create a ripple effect, impacting everything from school enrollments to local tax revenues. To foster a more sustainable model, a balanced approach is essential—one that supports renewable energy development while preserving the economic foundations of rural communities. This could include policies that prioritize renewable projects on less fertile lands or offer financial support to farmers who experience significant land-use changes, ensuring that renewable energy expansion does not come at the expense of rural prosperity and agricultural resilience.

Encroachment on Public Lands, National Parks, and Other Protected Areas

In addition to deserts and farmlands, large renewable energy projects often encroach on protected lands, including public lands, national parks, and wilderness areas. These regions are specifically set aside to safeguard biodiversity, conserve delicate ecosystems, and maintain the scenic beauty of natural landscapes. However, as demand for renewable energy rises, the pressure to develop these spaces has increased, with some proposed wind and solar projects aiming to establish installations near or even within close proximity to national parks and protected wilderness areas. This trend has sparked significant opposition from conservation groups, who argue that renewable energy development should not come at the cost of irreplaceable natural environments (Tomaschek & Miller, 2021).

The large physical footprint of wind and solar farms can be especially disruptive in these settings, often leading to habitat fragmentation that affects wildlife dependent on unbroken stretches of land for migration, breeding, and feeding. Species like desert bighorn sheep, pronghorn antelope, and migratory birds are particularly vulnerable, as the development of renewable projects fragments the continuous habitats they rely on, impacting movement patterns and increasing exposure to predators. Additionally, large energy infrastructure can alter the natural behavior of wildlife, displacing species sensitive to human activity or disrupting ecosystems that have been carefully managed for conservation.

The visual impact of wind turbines and expansive solar arrays also raises concerns, as these installations disrupt the scenic vistas that make national parks and protected lands so valuable to the public. Visitors come to these areas to experience untouched landscapes and pristine views, but the presence of industrial-scale infrastructure detracts from the aesthetic and recreational value of these spaces. For local communities that rely on eco-tourism, the encroachment of renewable energy projects into protected lands can have economic consequences, potentially deterring visitors and diminishing the revenue that supports local businesses.

To address these issues, a balanced approach is essential. Policymakers and energy companies must work collaboratively to prioritize the development of renewable energy on lands less critical to conservation, ensuring that the push for clean energy does not come at the expense of biodiversity and natural heritage. Conservation-minded planning, such as implementing buffer zones around protected areas and conducting comprehensive environmental assessments, can help mitigate the adverse impacts on these fragile ecosystems. By respecting the ecological and aesthetic importance of protected lands, it is possible to pursue renewable energy expansion while upholding the commitment to preserve nature's most cherished landscapes for future generations.

Conflicts with Conservation Goals and the Protection of Biodiversity

The expansion of renewable energy into protected lands creates significant conflicts with conservation goals, particularly concerning biodiversity protection. National parks, wildlife refuges, and wilderness areas are often home to rare and endangered species that rely on undisturbed environments to thrive. When these protected lands are opened for large-scale energy projects, the ecological consequences can be severe, potentially undoing decades of conservation efforts aimed at preserving these fragile ecosystems.

Wind turbines, for example, pose a well-documented threat to birds and bats, which are often killed by the turbine blades as they migrate or navigate through these areas. Species that are already vulnerable or have low reproductive rates are especially at risk, as even small population losses can have compounding effects on their survival. Solar farms present a different set of challenges, as they disrupt migration corridors and limit habitat availability for a range of animals. Species like pronghorn antelope, which depend on vast, open spaces, face barriers that impede their movement, making it difficult for them to access food, water, and breeding grounds. Similarly, reptiles and small mammals that rely on specific microhabitats can struggle to adapt when solar infrastructure changes the temperature, shade, and soil composition of their environment (Lovich & Ennen, 2017).

These projects can fragment habitats and create isolated patches of land that no longer support the natural behaviors of wildlife. The introduction of roads, fencing, and large equipment needed for construction further compounds habitat fragmentation, creating physical and psychological barriers for species sensitive to human activity. As a result, populations become isolated, reducing genetic diversity and making species more vulnerable to diseases, environmental changes, and climate shifts.

In addition to direct impacts on wildlife, renewable energy projects in protected lands can alter entire ecosystems. The removal of vegetation and disruption of soil during construction can lead to erosion, changes in water flow, and a reduction in native plant diversity. These alterations affect not only the species that depend on specific habitats but also the overall health and resilience of the ecosystem itself. Furthermore, the visual and noise pollution generated by wind and solar installations detracts from the natural experience that national parks and wilderness areas offer, potentially affecting tourism and the economic benefits that communities receive from conservation-focused areas.

To balance renewable energy development with conservation efforts, a strategic approach is needed that respects the ecological integrity of protected lands. Placing renewable projects on lands less critical to biodiversity or adopting stricter environmental assessments for protected areas can help mitigate adverse impacts. Ensuring that clean energy does not come at the cost of protected ecosystems is essential to uphold the long-term goals of both conservation and sustainable energy production, preserving biodiversity while advancing toward a sustainable future.

Difficulty in Reversing Land-Use Changes After Large Projects Are Decommissioned

One of the major challenges with large renewable energy projects is the irreversible nature of the land-use changes they impose. Once the infrastructure for a solar farm or wind turbines is established, restoring the land to its original, undisturbed condition is often difficult, if not impossible. The construction process alone involves extensive soil compaction from heavy machinery, which alters the soil structure and reduces its ability to retain water and nutrients. This compaction can take decades to naturally reverse, inhibiting plant growth and impacting the entire ecosystem (Pasqualetti, 2019).

Additionally, vegetation removed during construction may struggle to regrow, particularly in arid or semi-arid environments where plant recovery is slow and fragile ecosystems are especially vulnerable. Native plant species, some of which may be endangered or highly specialized, may fail to reestablish, leading to shifts in local biodiversity and habitat availability for wildlife. For desert and grassland ecosystems, which rely on specific plant species to stabilize soil and provide food and shelter for animals, this lack of recovery can lead to increased erosion, further degrading the land over time.

The environmental impact of these projects, therefore, extends well beyond their operational lifespan, creating long-term land degradation. When the renewable infrastructure eventually reaches the end of its life, the land may no longer support its original ecosystems, and any attempts to repurpose or remediate the site will be costly and complex. This reality raises essential questions about sustainable site selection, as well as the need for renewable projects to consider end-of-life land restoration strategies. By planning for the full lifecycle impact of these installations, it may be possible to mitigate some of the long-term environmental costs associated with large-scale renewable energy projects.

Soil Degradation, Erosion, and Other Lasting Environmental Impacts

Large energy projects can lead to significant soil degradation and erosion, especially in arid and semi-arid regions such as deserts and grasslands, where the soil is already vulnerable to disturbance. The construction and operation of solar and wind farms require extensive use of heavy machinery, which compacts and disrupts the soil, breaking up its structure and leaving it more susceptible to erosion by wind and water. This soil disturbance has cascading effects on the surrounding environment, beginning with the loss of vegetation. In ecosystems that rely on plant cover to stabilize the ground, the removal of vegetation opens the land to erosion, which further strips away nutrient-rich topsoil, hindering the regrowth of native plants (Lovich & Ennen, 2017).

The loss of soil structure also affects the land's ability to retain water, an especially critical function in dry regions where water availability is already limited. Compacted and eroded soils lose their capacity to absorb and store rainfall, which can lead to increased surface runoff and exacerbate drought conditions. This reduced water retention creates a hostile environment for native flora and fauna, further stressing ecosystems that depend on the scarce water resources. Additionally, disturbed areas are more prone to invasion by non-native species, which often outcompete native plants for resources and alter the local ecology. Invasive species can spread rapidly across degraded landscapes, reducing biodiversity and transforming the habitat in ways that make it inhospitable for native wildlife.

These impacts frequently persist long after the energy projects have ceased operation, as natural recovery in arid environments is slow and challenging. Without intervention, the land may take decades to restore, if it recovers at all, leaving behind a legacy of ecological degradation. The permanent changes to the landscape underscore the need for careful planning and impact assessments before establishing large-scale renewable projects. By considering soil preservation strategies and limiting disturbances to sensitive areas, energy developers can help mitigate the long-lasting environmental impacts of renewable installations in vulnerable ecosystems.

How Local Populations Are Fighting to Preserve Their Lands from Industrialization

Across the U.S., local populations are increasingly resisting the encroachment of large renewable energy projects on their lands. Communities in rural areas—where expansive wind and solar farms are often proposed—are voicing concerns over a range of issues, from the loss of valuable agricultural land to the visual impact of towering turbines and extensive solar arrays. Many rural residents fear these projects will alter the character of their landscapes, affecting not only the beauty of their surroundings but also the ecosystems that support local wildlife. Birds, bats, and other species native to these areas are often vulnerable to disruptions caused by large-scale renewable installations, adding another layer of concern for communities that feel a close connection to their natural environment.

These apprehensions have spurred grassroots organizing and legal challenges, with some communities successfully blocking projects altogether. Local opposition is often driven by a desire to preserve both the cultural and environmental integrity of their regions, which they feel is threatened by external, industrial-scale development. For example, in Vermont, a proposed large wind farm was halted due to organized resistance from residents, who cited concerns about the impact on local landscapes and wildlife (Warren & Birnie, 2020). In similar cases across the country, residents have lobbied for stricter regulations on renewable projects or advocated for community-scale renewable solutions that align more closely with local values and environmental needs.

The success of these movements highlights the need for a more inclusive approach to renewable energy planning—one that actively involves local communities and considers their unique concerns. By working collaboratively with residents and respecting the nuances of rural and ecological landscapes, developers and policymakers can foster renewable energy growth in ways that harmonize with local priorities. This approach not only promotes community buy-in but also ensures that the transition to clean energy respects the people and wildlife who call these regions home.

The Role of Community Engagement in Energy Project Planning

The rise of local resistance underscores the critical need for community engagement in the planning and development of renewable energy projects. Many large-scale renewable projects are initiated with limited input from the communities most directly impacted, leading to conflicts, delays, and, in some cases, project cancellations. This lack of consultation often results in communities feeling overlooked and their concerns unaddressed, which fosters distrust toward energy companies and policymakers alike. To mitigate these issues, energy developers and policymakers must prioritize early and meaningful community involvement, ensuring that residents' voices are incorporated into the decision-making process.

Engaging communities early allows developers to understand and address specific local concerns, whether they involve the loss of productive farmland, disruption to protected lands, or potential threats to wildlife. Offering more equitable financial arrangements, such as revenue-sharing models or community investment opportunities, can also build goodwill and help communities feel a sense of ownership over the projects in their areas. Additionally, considering alternative locations or scaling down project size to fit community preferences can reduce the visual and environmental footprint, preserving the character of rural and natural landscapes.

This collaborative approach not only helps to reduce opposition but also contributes to the overall success and sustainability of renewable energy projects. When communities are included as stakeholders and can trust that their needs are valued, they are more likely to support the project and participate in its long-term maintenance and success. By fostering a genuine partnership between developers and communities, renewable energy projects can advance in a way that balances the urgent need for clean energy with respect for local values and ecological integrity, creating a more harmonious and effective transition to sustainable energy sources (Tomaschek & Miller, 2021).

Alternative Energy Environmental and Economic Failures

In recent years, large-scale renewable energy projects have been promoted as a key part of the global shift toward cleaner energy. However, many of these projects have been plagued by inefficiencies, underperformance, and environmental damage. Despite massive government backing and taxpayer funding, numerous solar and wind energy projects have failed to deliver on their promises. Additionally, these projects often come with significant ecological costs, such as wildlife disruption and habitat destruction, raising questions about their overall sustainability and economic viability. Here are four examples of environmental and economic failures in alternative energy projects:

1. California: Ivanpah Solar Electric Generating System

The Ivanpah Solar Electric Generating System, located in California's Mojave Desert, was one of the most ambitious solar thermal projects, backed by $1.6 billion in federal loan guarantees. The project, however, encountered significant operational issues from the start, achieving only 40% of its projected energy output by 2014. One of the most severe issues was its impact on wildlife, especially birds, which were incinerated mid-flight by the intense heat produced by the solar mirrors. The project also relied on natural gas to supplement solar energy during periods of low sunlight, which further undermined its goal of producing clean energy. Despite massive investment, the project has been criticized for both its failure to meet energy targets and its negative environmental consequences, making it a cautionary tale about large-scale solar projects in fragile ecosystems.

2. Texas: Panhandle Wind Projects

Texas leads the U.S. in wind energy production, but several wind farms in the Texas Panhandle, such as the Brazos Wind Ranch and the Roscoe Wind Farm, have struggled with inconsistent energy production due to variable wind conditions. These projects also face significant transmission challenges, as the energy produced in these remote areas must travel long distances to urban centers, requiring costly infrastructure. Billions of dollars in federal tax credits and subsidies have been poured into these projects, but many have underperformed, leaving taxpayers to bear the burden. Additionally, the 2021 Texas winter storm exposed the limitations of wind energy's reliability during extreme weather events, further highlighting the challenges associated with large-scale wind farms.

3. **Wyoming: Chokecherry and Sierra Madre Wind Energy Project**

Located in Carbon County, Wyoming, the Chokecherry and Sierra Madre Wind Energy Project is one of the largest proposed wind farms in the U.S., expected to feature over 1,000 turbines. Despite receiving federal backing, the project has been delayed multiple times due to environmental concerns and legal challenges. The remote location of the project also raises concerns about the high cost of building the infrastructure needed to transmit the energy to more populated areas, which could require further taxpayer investments. Additionally, environmental groups have voiced concerns about the potential impact on wildlife, including birds and bats. The project remains in development but has already drawn criticism for its potential environmental damage, its high cost to taxpayers, and its uncertain future as an energy source.

4. **New York: Deepwater Wind (Now Orsted's South Fork Wind Project)**

Originally known as Deepwater Wind, the South Fork Wind Project off the coast of Long Island, New York, was celebrated as a groundbreaking offshore wind project. However, the project has been plagued by delays, cost overruns, and concerns over its impact on marine life, including whales and fishery stocks. Like many offshore wind projects, connecting the turbines to the mainland grid requires expensive taxpayer-funded upgrades, which have added to the overall cost of the project. The project has also faced significant opposition from local communities worried about the impact on their coastline and economy. Despite substantial taxpayer subsidies, the future of the project is in question due to its financial inefficiencies and unresolved environmental concerns.

Common Themes and Issues Across These Projects

Large-scale renewable energy projects often encounter a multitude of challenges, including inefficiency, underperformance, high taxpayer costs, and unintended environmental impacts. While these projects are driven by the goal of reducing carbon emissions, many fail to meet their projected energy output targets, necessitating additional taxpayer support to cover their escalating costs. For example, the massive initial investments in solar and wind farms frequently fail to yield proportional returns in energy generation, resulting in long payback periods that strain both public resources and investor confidence. The reliance on public funding to sustain these projects raises questions about fiscal responsibility, as taxpayers ultimately bear the burden of supporting projects that may not deliver on their promises.

Despite their environmental intentions, large renewable projects often produce unintended ecological consequences. The construction and operation of these sites can disrupt fragile ecosystems, displacing native species and damaging biodiversity. Habitats for endangered species, particularly in sensitive regions like deserts and grasslands, are often altered or destroyed, and migration patterns for wildlife are affected as once-pristine lands are converted into industrial sites. Additionally, wind farms pose direct threats to birds and bats, while large solar farms disrupt local flora and can contribute to soil erosion and degradation. These ecological costs can undermine the environmental goals of renewable projects, diminishing their benefits and potentially causing long-term damage to the landscapes they were intended to protect.

One significant challenge of large renewable energy installations is their remote locations, which are often chosen to capitalize on optimal sunlight or wind conditions. However, placing projects far from urban centers introduces costly infrastructure requirements, such as high-voltage transmission lines, to carry the energy generated to areas where demand is highest. The expense and energy loss involved in long-distance transmission reduce overall efficiency, detracting from the projects' cost-effectiveness and contributing to the cumulative environmental impact through additional land use for power lines and substations.

While large-scale renewable energy projects are frequently seen as central to the clean energy transition, their economic inefficiencies, high environmental costs, and dependency on taxpayer funding raise concerns about their long-term viability. These projects illustrate the need for more sustainable, decentralized alternatives that better balance the goals of clean energy production with ecosystem preservation and prudent use of taxpayer resources. Smaller, community-centered renewable solutions, such as rooftop solar and local wind installations, offer a more adaptable approach, reducing the need for extensive infrastructure and preserving natural and agricultural lands from industrial-scale disruption.

The impact of large renewable energy projects on America's landscapes is both profound and far-reaching. From deserts to farmlands, the economic and environmental costs of these installations often outweigh their projected benefits. While renewable energy is crucial to reducing carbon emissions, it is essential to consider the full spectrum of consequences that accompany industrial-scale projects. As communities across the nation increasingly resist the encroachment of these large installations, there is a growing call for renewable energy approaches that prioritize sustainability, minimize environmental disruption, and foster local engagement. By reimagining renewable energy development with a focus on smaller-scale, community-driven projects, the U.S. can pursue a cleaner energy future that respects and protects its diverse landscapes and ecosystems.

Chapter 6

Ivanpah Illusion vs. Rooftop Reality
A Comparative Study of Large-Scale Solar vs. Distributed Rooftop Solar

In the pursuit of renewable energy, the Ivanpah Solar Electric Generating System stands as a prominent example of large-scale concentrated solar power (CSP). Located in California's Mojave Desert, Ivanpah occupies approximately 3,500 acres and has a capacity of 392 megawatts (MW), theoretically enough to power over 100,000 homes (BrightSource Energy, 2021). Ivanpah's design and sheer scale initially sparked excitement, as it was one of the largest and most ambitious CSP projects worldwide, relying on vast fields of mirrors to concentrate sunlight onto towers that generate steam, powering turbines to produce electricity. However, despite this impressive scale, Ivanpah has faced numerous challenges that have limited its effectiveness as a reliable renewable energy source (California Energy Commission, 2023).

The plant has consistently fallen short of its output targets, generating less power than initially projected due to issues such as weather variability, unexpected maintenance needs, and significant dependence on natural gas for daily start-up operations (U.S. Department of Energy [DOE], 2022). Additionally, Ivanpah's heliostats—its thousands of mirrors—have encountered alignment and malfunction issues, which reduce their ability to focus sunlight precisely on the tower, limiting power output and increasing maintenance costs (BrightSource Energy, 2021). Complicating matters, the site struggles with vegetation overgrowth that obstructs the heliostats, adding an unexpected layer of upkeep to prevent weeds from affecting energy production (California Energy Commission, 2023). To address the intermittent sunlight that CSP systems rely on, Ivanpah's natural gas use for startup has surpassed initial projections, to the extent that some years it operates almost like a small gas plant, further undermining its green credentials and adding to its carbon footprint (U.S. Department of Energy [DOE], 2022).

These shortfalls undermine its intended environmental benefits, as burning natural gas to jumpstart the facility contributes to carbon emissions, counteracting Ivanpah's renewable goals. Moreover, Ivanpah's high operational costs have been a burden, with expenses outpacing revenue as it struggles to meet energy production expectations (BrightSource Energy, 2021). This high-cost structure poses ongoing concerns about the sustainability of large-scale CSP projects, especially when compared to more flexible and scalable alternatives.

Environmental impacts further complicate Ivanpah's operations. Situated in the Mojave Desert, the facility disrupts the delicate desert ecosystem, affecting native wildlife, including the threatened desert tortoise. The intense heat produced by its mirrors also poses hazards to birds, which are drawn to the reflective surfaces and suffer fatal burns upon entering concentrated solar beams (Defenders of Wildlife, 2022). These ecological impacts have brought Ivanpah under scrutiny from conservationists and raised concerns about the broader viability of CSP in sensitive habitats (California Energy Commission, 2023).

In comparison, rooftop solar installations offer a more decentralized and efficient approach that minimizes environmental impact while delivering local, reliable power. By generating power at the point of use, rooftop systems reduce transmission losses and avoid the extensive land use and ecological footprint of centralized solar farms (National Renewable Energy Laboratory [NREL], 2022).

Rooftop solar installations also empower individual homeowners and communities, allowing them to harness clean energy without the high maintenance costs and environmental disruptions associated with projects like Ivanpah. As a scalable, adaptable solution, rooftop solar demonstrates the potential for renewable energy to be both impactful and sustainable without the burdens associated with large-scale CSP projects (Solar Energy Industries Association [SEIA], 2021).

Ivanpah's Energy Production vs. Rooftop Solar's Potential

Ivanpah's design was ambitious, yet it relies heavily on consistent sunlight and weather patterns, causing significant fluctuations in output. Unlike photovoltaic (PV) systems, which convert sunlight directly into electricity, Ivanpah uses a concentrated solar power (CSP) system that focuses sunlight onto a central tower to produce steam, which drives a turbine to generate electricity. This process is not only complex but also sensitive to cloud cover and variations in sunlight, leading to unpredictable production levels. For instance, due to the intermittency of sunlight and the need for a natural gas backup, Ivanpah's real-world capacity often falls short of its 392 MW design potential, affecting both reliability and environmental performance (U.S. Department of Energy [DOE], 2022). While Ivanpah's location in the Mojave Desert allows for substantial sunlight exposure, it is also far from many populated areas, leading to energy losses during transmission, as electricity must travel long distances before reaching consumers (California Energy Commission, 2023).

By comparison (see below table), if 100,000 homes each installed a 5-kilowatt (kW) rooftop solar system, the collective generation capacity would reach 500 MW, surpassing Ivanpah's theoretical output without requiring vast, ecologically sensitive land (National Renewable Energy Laboratory [NREL], 2022). This decentralized setup brings production closer to the point of consumption, reducing transmission losses and enhancing overall efficiency. Unlike CSP systems, rooftop PV systems generate electricity directly on-site, meaning homes can produce power where it's needed without the complex infrastructure or transmission lines associated with centralized solar plants, thus minimizing energy waste and improving overall sustainability (Solar Energy Industries Association [SEIA], 2021).

Aspect	Ivanpah Solar Plant	Rooftop Solar (Distributed)
Energy Capacity	392 MW (designed capacity)	500 MW (achieved with 100,000 homes each with a 5 kW system)
Land Use	3,500 acres in Mojave Desert	No additional land needed; uses existing rooftops
Environmental Impact	Significant land disruption; impacts desert wildlife, especially the desert tortoise; risks to bird populations from reflective mirrors (BrightSource Energy, 2021; Defenders of Wildlife, 2022)	Minimal impact, no habitat disruption, low risk to wildlife
Transmission Loss	High, due to distance from urban centers	Low, as energy is generated near point of consumption
Intermittency and Reliability	Dependent on sunlight, with natural gas backup for consistency (DOE, 2022)	Can be supplemented with battery storage for reliability
Economic Impact	High initial and ongoing costs; funded largely by taxpayer subsidies	Direct savings to homeowners; incentivized by net metering (SEIA, 2021)
Community Resilience	Centralized, less resilient to grid outages	Decentralized, increases resilience and reduces grid dependency
Scalability	Limited by available land and infrastructure costs	Highly scalable with individual rooftop installations

Environmental Impact: Land Use and Wildlife Disruption

Ivanpah's extensive land use poses a substantial environmental challenge. Large solar farms like Ivanpah disrupt native ecosystems and displace wildlife, including the threatened desert tortoise, whose habitat has been significantly affected (Defenders of Wildlife, 2022). The construction and operation of Ivanpah have raised concerns about its long-term impacts on biodiversity in the Mojave Desert (California Energy Commission, 2023). Additionally, the facility's reflective mirrors pose risks to bird populations, as birds are often attracted to the light, mistaking it for water, leading to fatal collisions (BrightSource Energy, 2021).

In contrast, rooftop solar requires no additional land and leverages existing structures, making it an inherently low-impact solution. By generating energy directly on rooftops, it avoids the need for new infrastructure or habitat clearance, preserving natural landscapes and reducing ecological strain (EnergySage, 2021). Moreover, rooftop systems do not pose the same risks to wildlife as large solar arrays, making them a more environmentally responsible option (SEIA, 2021).

Economic and Community Benefits of Rooftop Solar

The economic benefits of rooftop solar installations extend well beyond individual households, offering community-wide advantages that contrast sharply with the costly nature of large-scale renewable energy projects. Unlike industrial wind and solar farms, which often require significant taxpayer subsidies to cover their high infrastructure and operational costs, rooftop solar installations deliver direct financial savings to homeowners. These systems not only reduce monthly electricity bills but can also increase property values, making homes with rooftop solar more attractive in the real estate market (NREL, 2022). Programs like net metering further incentivize adoption, allowing homeowners to earn credits for any excess energy they send back to the grid. This arrangement makes rooftop solar a sustainable, economically viable investment that provides immediate and long-term returns for communities (SEIA, 2021).

Moreover, rooftop solar promotes energy resilience by decentralizing energy production, which reduces dependency on large, centralized grids. When households generate their own electricity, they are less affected by grid disruptions and blackouts, particularly when rooftop solar is paired with battery storage systems that provide backup power. This added resilience is especially crucial in regions vulnerable to extreme weather events, such as hurricanes or wildfires, where the reliability of centralized grids can be compromised. In these situations, rooftop solar systems with battery storage can keep homes powered independently, enhancing community stability and security during emergencies (EnergySage, 2021).

By fostering energy independence at the local level, rooftop solar also mitigates the need for extensive transmission infrastructure, which is both costly and environmentally intrusive. As rooftop solar becomes more widespread, communities can reduce their reliance on long-distance transmission lines and centralized power plants, cutting down on transmission losses and reducing the overall strain on the power grid. This localized approach to energy generation not only benefits homeowners but also contributes to the resilience and efficiency of the entire energy system, creating a cleaner, more adaptable, and more economically sustainable energy landscape.

The Case for Rooftop Solar Over Large-Scale Projects

While Ivanpah represents an ambitious effort to harness renewable energy on a large scale, its challenges highlight the limitations of centralized solar farms. By contrast, rooftop solar offers a compelling, decentralized alternative that delivers higher efficiency, reduced environmental impact, and enhanced community resilience. As renewable energy initiatives continue to expand, prioritizing rooftop solar over sprawling desert installations could provide a more sustainable and scalable path to a cleaner energy future, one that respects natural landscapes and integrates seamlessly into urban settings.

The energy production comparison between Ivanpah and an equivalent number of homes equipped with rooftop solar underscores the efficiency of decentralized solar power. Outfitting 100,000 homes with 5-kilowatt (kW) rooftop systems would generate 500 MW of power, a full 28% more than Ivanpah's 392 MW capacity. This increased output is achieved without consuming vast tracts of land or requiring complex transmission infrastructure, as rooftop systems generate power directly where it is used. Avoiding the need for long-distance transmission also reduces energy losses, which are inherent in centralized systems like Ivanpah that must transport power from remote locations to urban centers. By generating electricity locally, rooftop solar maximizes the efficiency of each kilowatt produced, further underscoring its economic and environmental advantages.

A shift toward community-based rooftop installations not only boosts energy output but also enhances resilience and energy independence at the local level. In a decentralized network, power generation is dispersed across thousands of homes, creating a stable and locally controlled grid that is less vulnerable to large-scale outages. This resilience is increasingly critical as climate-related extreme weather events become more common, with localized solar systems providing reliable power even when centralized grids face disruptions. Additionally, rooftop solar allows communities to take an active role in renewable energy adoption, fostering local investment and creating economic opportunities within the community itself.

The decentralized approach thus represents a transformative shift in renewable energy, yielding higher efficiency, reduced environmental impact, and enhanced community resilience. By focusing on scalable, community-centered rooftop installations, renewable energy efforts can meet sustainability goals more effectively, creating a cleaner, more resilient energy future that aligns with both environmental priorities and community needs.

Chapter 7

Rooftop Solar: An Untapped Revolution

This chapter explores the transformative potential of a technology that has yet to be fully realized. While much of the focus in renewable energy has been on large, centralized projects, the true revolution may lie above our heads—on rooftops across America. Rooftop solar offers an unparalleled opportunity to decentralize energy production, reduce electricity costs, and empower individuals and communities to generate their own clean power. Despite its immense potential, rooftop solar remains an underutilized resource in the national energy strategy. This chapter delves into the untapped promise of rooftop solar, examining how it could redefine the future of energy, democratize power generation, and provide a more efficient and sustainable solution to America's energy needs.

Overview of Rooftop Solar Technology

Rooftop solar technology has fundamentally changed how electricity is generated, providing a decentralized and efficient solution that uses previously unused roof space. Photovoltaic (PV) panels work by converting sunlight directly into electricity, meeting the energy demands of individual homes or businesses. The technology has advanced rapidly, with modern PV panels becoming more efficient and cost-effective. These advancements make rooftop solar viable even in regions with lower direct sunlight, as panels now generate more electricity from less sunlight.

In addition to producing electricity, rooftop solar can be integrated with energy storage systems like batteries, which store excess energy generated during the day for use at night or during periods of low sunlight. This ability to store energy significantly increases the utility and reliability of solar power. Homeowners and businesses can rely less on the grid, reducing electricity bills and increasing energy independence, particularly during peak demand or grid outages.

The appeal of rooftop solar lies in its combination of efficiency and adaptability. Solar panels on residential and commercial rooftops take advantage of unused surfaces, avoiding the need for additional land or significant new infrastructure. This decentralized energy model bypasses many of the inefficiencies of large-scale projects, such as energy loss during long-distance transmission, and offers a more flexible solution to meet localized energy needs. Moreover, rooftop solar contributes to grid resilience by producing energy closer to where it is consumed, reducing the strain on public utilities (IRENA, 2020).

As rooftop solar becomes more accessible and affordable, thanks to advances in PV technology, it is increasingly seen as a key element in the global transition to clean energy. With higher efficiency rates, modern solar panels are better suited for diverse geographic regions, even those with less direct sunlight. This broadens the potential for solar energy adoption, making rooftop solar an attractive option for urban and rural areas alike. The inclusion of battery storage options further enhances its value, allowing for energy self-sufficiency and mitigating the need for backup power from the grid. This shift towards sustainable, on-demand energy places rooftop solar at the forefront of renewable energy solutions for households and businesses.

The compactness and affordability of modern solar systems also allow them to be installed in areas where traditional energy projects might not be feasible, making it a highly scalable solution. This scalability, coupled with declining costs and increasing government incentives, positions rooftop solar as a leading player in the renewable energy revolution, poised to reduce dependence on fossil fuels while promoting energy security.

How Rooftop Solar Can Meet Localized Energy Needs More Efficiently

Unlike large-scale renewable projects that require vast tracts of land and infrastructure to transmit electricity, rooftop solar operates on a highly localized level, meeting energy needs directly at the point of generation. This decentralized model means that energy produced by rooftop panels is used by the same building or neighborhood, avoiding the energy losses that occur when electricity is transmitted over long distances from remote solar farms (Baker, 2022). These transmission losses, which can account for 5-10% of energy in large-scale projects, are virtually eliminated with rooftop solar.

Moreover, rooftop solar is adaptable, allowing homeowners and businesses to tailor the size of their system based on specific energy consumption patterns. For instance, a small household can install a modest system, while a larger business can invest in a more extensive setup. This flexibility is a key advantage in densely populated urban environments, where space for large-scale renewable projects is limited but energy demand is high. Rooftop solar utilizes existing structures without the need for additional land, making it an ideal solution for cities where expansion is constrained.

By generating energy on-site, rooftop solar also improves grid resilience, reducing the strain on public utilities and enhancing the reliability of local power supplies. In times of grid outages or peak demand, homes and businesses with rooftop solar, especially when coupled with battery storage, can maintain a steady energy supply, making the grid less vulnerable to disruptions. This capability is particularly important during extreme weather events or other emergencies when centralized energy systems may be compromised.

Additionally, rooftop solar supports energy independence and sustainability goals. The decentralized nature of these systems ensures that power is produced and consumed locally, empowering consumers to manage their energy production and consumption more directly. Over time, this reduces reliance on large utility companies, fosters energy autonomy, and drives down long-term energy costs, while simultaneously contributing to a greener and more resilient energy grid

Financial Incentives, Net Metering, and the Economics of Home Solar Solutions

One of the most significant factors driving the widespread adoption of rooftop solar is the availability of financial incentives. Governments at all levels—federal, state, and local—have introduced policies such as subsidies, tax credits, and rebates to make solar energy more affordable. The U.S. federal solar investment tax credit (ITC), for instance, allows homeowners to deduct a percentage of their solar installation costs from their federal taxes, significantly reducing the financial barrier to going solar. This credit covers 30% of installation costs, making it one of the most impactful incentives available (Solar Energy Industries Association, 2023).

At the state level, policies such as Net Energy Metering (NEM) have become critical for homeowners who install solar systems. Net metering allows homeowners to feed excess electricity generated by their rooftop solar systems back into the grid, receiving credits on their electricity bills in return. This system works particularly well during peak sunlight hours, when solar panels often generate more electricity than a home needs. These credits can then be used during periods of lower energy production, such as at night or on cloudy days. As a result, net metering effectively allows homeowners to bank energy credits, improving the overall return on investment and lowering long-term energy costs.

In addition to financial incentives and net metering, the falling cost of solar panels has also contributed to the increased adoption of rooftop solar. Solar panel prices have decreased dramatically over the past decade due to technological advancements and economies of scale in production. Combined with government incentives, this decline in price makes it feasible for more homeowners to invest in solar. In fact, most homeowners can expect a return on investment (ROI) within 6-10 years, depending on local utility rates and incentive programs (National Renewable Energy Laboratory [NREL], 2021). After the system pays for itself, homeowners enjoy decades of savings, as solar panels typically last 25-30 years.

Moreover, with the ability to eliminate or significantly reduce electricity bills, rooftop solar presents an opportunity for long-term financial security. By generating their own electricity, homeowners protect themselves from rising energy prices, while also contributing to a more sustainable energy future. This makes rooftop solar not only a smart economic choice but also a key player in reducing carbon footprints, enhancing energy independence, and promoting environmental sustainability.

Rooftop solar represents a highly impactful yet underutilized tool in the transition toward renewable energy. It provides localized, on-site energy generation, eliminating many of the inefficiencies seen in large-scale projects, such as energy loss during transmission. With the added benefits of financial incentives, such as tax credits and net metering, rooftop solar becomes a viable and sustainable alternative for both homeowners and businesses. As technology advances and costs continue to decline, rooftop solar adoption is expected to play a critical role in shaping America's clean energy future, offering affordability, efficiency, and energy independence.

What makes rooftop solar particularly promising is its scalability and adaptability to various environments. It allows individual buildings to produce their own electricity, directly reducing reliance on traditional power grids. By generating energy where it is consumed, rooftop solar not only supports energy independence but also reduces strain on centralized power infrastructure, increasing overall grid resilience. Furthermore, rooftop solar has the flexibility to cater to specific energy demands, whether in residential homes, commercial properties, or industrial facilities, making it highly adaptable in both urban and rural settings.

As more homeowners and businesses realize the economic and environmental advantages of rooftop solar, its role in the broader clean energy strategy will continue to expand. Financial incentives, coupled with technological innovations such as improved photovoltaic (PV) efficiency and affordable energy storage solutions, will likely drive even greater adoption rates. Rooftop solar offers a path to a more decentralized and democratized energy system, one in which individuals and communities have greater control over their energy production, costs, and carbon footprint.

Rooftop solar remains highly underutilized, in part because corporate interests have historically benefited from massive subsidies for large-scale renewable energy projects. These corporations, accustomed to receiving significant financial backing from federal and state governments, have less incentive to promote decentralized, homeowner-based energy solutions like rooftop solar. Unlike utility-scale projects, which garner corporate profits and subsidies, rooftop solar empowers individuals and communities to generate their own electricity, reducing the influence of large energy companies. This shift in power dynamics has slowed the widespread adoption of rooftop solar, despite its clear benefits.

Ultimately, rooftop solar is poised to become a transformative cornerstone of America's renewable energy landscape, providing a sustainable and cost-effective solution to both local and national energy needs. Its ability to generate power directly where it's consumed offers unmatched efficiency, while financial incentives and decreasing installation costs make it increasingly accessible for homeowners and businesses alike. However, its adoption has been slowed by a system that favors large, corporate-backed projects, which receive substantial subsidies and policy support. This imbalance has often left rooftop solar underfunded and underutilized, despite its clear advantages. As awareness grows and technology continues to improve, rooftop solar will empower communities, reduce dependence on corporate-controlled utilities, and drive a cleaner, more decentralized energy future, offering a resilient alternative to traditional grid systems.

Chapter 8

Decentralization
The Future of Energy

Let's envisions a world where power generation is no longer confined to massive, centralized plants but instead distributed across millions of homes, businesses, and communities. In this future, energy is produced closer to where it is consumed, reducing inefficiencies and increasing resilience. Decentralized systems, such as rooftop solar and community microgrids, offer a flexible, sustainable alternative to the aging infrastructure of today. As technology advances and the need for climate action grows, decentralization is emerging as the key to a cleaner, more reliable, and more equitable energy future. This chapter explores how shifting away from centralized models could transform the energy landscape, putting power—literally and figuratively—into the hands of consumers and communities

The Advantages of Decentralizing Energy Production

Decentralizing energy production represents a transformative shift in how power is generated, consumed, and distributed, challenging traditional models and reshaping the energy landscape. Conventional energy systems rely heavily on large, centralized power plants—whether fossil-fuel, nuclear, or hydroelectric—that generate electricity and distribute it across vast distances via intricate grid networks. While this model offers efficiency at a large scale, it comes with significant downsides, including vulnerability to widespread outages, inefficiencies in energy transmission, and high operational and maintenance costs. For example, centralized grids lose a percentage of the generated power due to resistance in transmission lines, especially over long distances, which is both wasteful and costly. In contrast, decentralized energy production, achieved through rooftop solar, small-scale wind turbines, and other localized energy sources, brings power generation closer to the point of consumption, substantially reducing energy loss and increasing overall system efficiency (Carley & Konisky, 2020).

Decentralized energy production empowers communities and individuals to have greater control over their energy supply, creating a more resilient and self-sufficient model. Instead of depending solely on distant power plants, consumers can generate their own electricity and contribute excess energy back to the grid or store it for later use. This local ownership of energy encourages greater energy independence and security, enabling households and communities to better withstand disruptions to the main grid—whether due to extreme weather, technical failures, or supply shortages. In the event of localized power outages, decentralized systems can often continue operating, maintaining electricity access for critical functions. Moreover, with the integration of energy storage systems, like home batteries, and advanced grid technologies, communities can manage their energy use more dynamically, smoothing out supply fluctuations and reducing peak demand pressures on the main grid.

Decentralizing energy production also allows for a more flexible and adaptable energy infrastructure. Smaller, localized power systems are easier to scale and integrate with emerging technologies, such as smart grid solutions and automated energy management systems, which can optimize energy flow and storage based on real-time data. This adaptability makes it simpler to incorporate renewable sources incrementally, aligning energy supply with the specific needs and growth rates of local populations. For example, smart grid technologies can help balance supply and demand across a decentralized network, automatically shifting power from high-supply to high-demand areas to maintain grid stability without requiring large-scale, high-cost interventions.

Additionally, decentralization fosters healthy competition in the energy market, challenging the monopoly that large utility companies traditionally hold. As more individuals and communities adopt localized energy solutions, the market opens up for innovation in renewable technologies, energy storage, and efficiency improvements. This competition can drive down costs, incentivize technological advancements, and encourage the development of sustainable energy solutions that are tailored to the needs of diverse regions. Localized power generation also encourages environmentally responsible choices, as communities have a vested interest in reducing pollution and preserving local landscapes, leading to increased support for clean, renewable energy sources.

In essence, decentralized energy production represents a more resilient, efficient, and democratized approach to power generation. It leverages the unique strengths of local communities, promotes sustainability, and challenges the traditional energy monopoly. By empowering individuals and communities to contribute directly to the energy supply, decentralization fosters a cleaner, more adaptable, and innovative energy future.

Energy Independence at the Household Level

One of the most significant benefits of decentralized energy is the energy independence it provides at the household level. Homeowners and businesses equipped with rooftop solar panels or small wind turbines can generate much, if not all, of their electricity needs on-site. This reduces dependence on utility companies, protecting individuals from rising energy prices and insulating them from grid failures. The economic appeal is clear: after the initial investment in a rooftop solar system, many households can significantly reduce or even eliminate monthly electricity bills (NREL, 2021).

Energy independence gives households the power to control both their energy production and consumption. This autonomy becomes even more valuable with the integration of energy storage systems, like batteries. Excess energy produced during the day can be stored and used at night or during periods of low sunlight, ensuring a continuous supply. For many households, this means minimal reliance on the grid, allowing them to maintain a steady flow of energy even when external conditions—such as weather or infrastructure issues—compromise the central power grid.

In regions that are prone to frequent blackouts, such as those caused by hurricanes, wildfires, or other natural disasters, energy independence can be critical. Areas with unreliable grid infrastructure often face extended power outages during storms or peak demand periods, leading to widespread discomfort, economic loss, and health risks. Households with their own energy generation and storage systems, however, can continue to power their homes, refrigerators, medical devices, and communication systems during these critical times, providing a layer of energy security and peace of mind.

Moreover, for rural and remote areas where grid connections are weak or inconsistent, the ability to generate electricity on-site is a game-changer. Rooftop solar and decentralized energy systems can enable communities that previously lacked reliable electricity to take control of their energy futures, reducing their dependence on external sources and improving their overall quality of life. As decentralized energy systems like rooftop solar become more affordable and accessible, the potential for widespread energy independence will likely transform how individuals and communities approach their energy needs.

Resilience in the Face of Grid Failures and Climate Disasters

As climate disasters become more frequent and severe, ensuring energy resilience has emerged as a critical concern. Traditional, centralized power grids are highly vulnerable to natural disasters like hurricanes, wildfires, and extreme weather events. Hurricanes in the southeastern U.S., for example, have caused massive grid failures, leaving millions without power for extended periods. In contrast, decentralized energy systems, such as rooftop solar with battery storage, provide resilience by allowing homes and businesses to maintain electricity even if the main grid fails.

The advantage of decentralized systems lies in their localized energy generation. Instead of relying on a single, large power plant that could be disabled during a disaster, decentralized systems distribute energy production across numerous smaller installations. This diversification reduces the risk of widespread outages. For instance, in areas like California, where wildfires frequently threaten grid infrastructure, decentralized energy ensures that individual homes and businesses can remain operational. Similarly, in hurricane-prone regions like the Gulf states, homes equipped with solar and storage systems can maintain electricity when central power plants are offline.

Decentralized systems not only increase resilience but also speed up recovery after grid failures. In many instances, communities with decentralized infrastructure can restore power much faster than those dependent on centralized grids, which may take days or even weeks to repair. For example, after hurricanes, decentralized energy systems enable critical functions—such as communication, refrigeration, and medical devices—to continue operating, significantly alleviating the burden on emergency responders.

This ability to provide localized, reliable power during crises underscores the value of decentralized energy in protecting households and businesses from the impacts of climate disasters. As extreme weather becomes more common, decentralized energy systems like rooftop solar will play an increasingly vital role in creating a resilient and secure energy future.

Decentralization represents a transformative shift in the future of energy, delivering unparalleled benefits in terms of efficiency, independence, and resilience. By generating electricity at the point of consumption, systems like rooftop solar dramatically reduce energy loss, reliance on vulnerable grids, and susceptibility to outages. For households, it provides autonomy over energy costs and security; for communities, it strengthens resilience in the face of intensifying climate disasters. As technological advancements drive down costs and enhance performance, decentralized energy will become a cornerstone of a sustainable, secure, and climate-resilient future. The evolution of this model is not just inevitable—it is essential for a world increasingly affected by environmental uncertainty and rising energy demands. The path to a cleaner, more secure energy future lies in empowering individuals and communities through decentralized solutions that redefine how energy is produced and consumed.

Chapter 9

Policy Failures and Misdirected Subsidies

In the transition to renewable energy, government policies have played a crucial role in shaping the current landscape. However, many of these policies have disproportionately favored large-scale energy projects over smaller, more efficient solutions like rooftop solar. Through substantial subsidies and the influence of corporate lobbying, the focus has remained on centralized energy models, often at the expense of localized systems that offer greater efficiency, sustainability, and resilience. This chapter explores the policy failures that have misdirected resources and the potential for change to support decentralized energy solutions, particularly rooftop solar.

The Role of Government in Promoting Large-Scale Projects Over Smaller, More Efficient Ones

Government policy has historically prioritized large-scale energy projects, offering substantial subsidies and incentives to centralized power plants like utility-scale wind and solar farms. While these large projects have contributed to renewable energy growth, they often fall short in efficiency compared to decentralized solutions such as rooftop solar. This preference stems from long-standing policies that reward high-capacity energy generation, overlooking the potential of smaller, localized systems. As a result, energy investments have skewed toward centralized grids, leaving decentralized models underfunded and underutilized (Carley & Konisky, 2020).

Large-scale projects often face significant inefficiencies, including energy loss during long-distance transmission and extensive land-use impacts. Even though they provide substantial amounts of energy to the grid, they are less efficient in terms of localized energy production. Despite these drawbacks, these projects have consistently received the lion's share of government support through grants, tax credits, and other incentives. The focus on centralized systems stifles the adoption of smaller-scale, decentralized energy systems like rooftop solar, which offer better grid resilience and can meet community-specific needs more efficiently.

Decentralized energy production, such as rooftop solar, generates electricity closer to the point of consumption, reducing transmission losses and land-use impact. However, the heavy government focus on large-scale renewable projects has inadvertently overlooked this potential, leaving rooftop solar and other decentralized models underfunded and less incentivized. This has resulted in an energy system that prioritizes large, centralized grids, rather than exploring the broader benefits of decentralizing energy production, which could improve sustainability, energy independence, and resilience.

By promoting large-scale projects, government policies have overlooked the transformative power of decentralization. Decentralized solutions like rooftop solar can more efficiently meet the energy needs of individuals and communities, with minimal environmental and transmission-related inefficiencies. To foster a more sustainable future, policymakers need to shift their focus toward decentralization and provide the necessary incentives and support to scale up smaller, localized energy production systems.

How Subsidies and Lobbying Have Shaped the Current Energy Landscape

The current energy landscape has been heavily influenced by subsidies and lobbying efforts from powerful corporations and established energy interests. Large energy companies, particularly those involved in centralized, utility-scale projects, have long benefited from significant government subsidies. These companies have leveraged their political connections and lobbying power to secure billions of dollars in taxpayer-funded subsidies for large-scale projects, including massive solar farms, wind farms, and even fossil fuel-based energy generation. This entrenched system of support has created a policy environment that disproportionately favors large, centralized energy projects, often at the expense of smaller, decentralized systems like rooftop solar.

Through aggressive lobbying efforts, these energy corporations have maintained a policy structure that protects their market dominance. One of the ways they have done this is by actively opposing policies that promote decentralization, as these policies threaten their control over the energy market. Lobbying against net metering, for example, is a common tactic employed by large utility companies. Net metering allows owners of rooftop solar systems to sell excess electricity back to the grid, reducing their overall energy costs and contributing to local energy resilience. By lobbying to restrict net metering, utility companies seek to protect their revenue streams and maintain their control over energy production and distribution, effectively limiting the potential growth of rooftop solar and other small-scale renewable energy technologies.

In addition to influencing policy, these lobbying efforts shape public perception of renewable energy. Large-scale projects are often marketed as essential to achieving national energy goals, while decentralized solutions like rooftop solar are sidelined. The public is often led to believe that utility-scale projects are the only viable path forward, even though they come with significant inefficiencies and environmental impacts, such as energy loss during transmission and large land-use requirements. This narrative allows utility companies to maintain their grip on the market, perpetuating an energy system that prioritizes large-scale infrastructure and centralized control over more efficient and adaptable local solutions (Joskow, 2019).

This dominance of large energy interests has led to an imbalanced energy system, where smaller, more sustainable technologies struggle to gain traction despite their potential to reshape the energy landscape. Rooftop solar and other decentralized systems offer numerous benefits, such as improved energy efficiency, greater grid resilience, and localized energy independence. However, because they challenge the existing energy structure, they face significant policy and market barriers. The current landscape, shaped by lobbying and political influence, continues to favor outdated models of energy production that are less sustainable in the long term.

As the demand for cleaner, more resilient energy grows, there is an urgent need for policy reform that levels the playing field for decentralized solutions. Without such reforms, the energy system will remain heavily skewed toward large corporate interests, stifling the innovation and widespread adoption of sustainable, decentralized energy technologies like rooftop solar.

The Potential for Policy Changes to Support Rooftop Solar

While government policies have historically supported large-scale energy projects, there is significant potential for policy changes that could bolster the adoption of rooftop solar. Shifting subsidies and incentives toward decentralized energy production would promote more efficient, resilient, and sustainable systems.

One of the most impactful changes would involve expanding net metering programs. Net metering allows homeowners to sell excess energy back to the grid, reducing their energy bills and creating a financial incentive for more people to adopt rooftop solar. Expanding these programs and ensuring their long-term stability would encourage greater investment, as the financial benefits would become clearer and more reliable. A strong, uniform net metering policy at both state and federal levels could remove the uncertainty that currently discourages many potential solar adopters.

Additionally, offering direct subsidies and tax credits targeted at small-scale energy producers—rather than focusing exclusively on utility-scale projects—could drive a decentralized energy revolution. Increased financial support for rooftop solar and energy storage systems would make it more accessible, especially for low- and middle-income households. This democratization of solar energy could help create a more equitable energy landscape, empowering individuals and communities to take part in the renewable energy transition.

Governments can also implement policies that simplify the permitting process for rooftop solar installations. Currently, bureaucratic barriers and complex permitting processes slow the adoption of rooftop solar, making it difficult for individuals and small businesses to navigate the system. Streamlining regulations and standardizing procedures across states and municipalities would lower costs, reduce wait times, and make it easier for more people to install solar systems.

Finally, community solar programs present another important policy tool for expanding access to rooftop solar. These programs allow groups of people—such as renters, homeowners with unsuitable roofs, or those living in multi-unit buildings—to share the benefits of a solar energy system. By pooling resources and sharing energy generation across multiple properties, community solar makes decentralized energy more accessible to a broader population, further supporting the shift toward a sustainable and equitable energy future.

With these policy shifts, governments could transform the energy landscape, shifting from large, centralized projects to decentralized, community-driven solutions that offer better resilience, efficiency, and long-term sustainability.

For far too long, government policies have favored large, centralized energy projects, often dictated by corporate interests and sustained through extensive subsidies, tax credits, and lobbying efforts. This approach has not only burdened the American taxpayer but has also inflicted significant environmental costs. While job creation remains critical, it should not come at the expense of long-term sustainability and equitable energy solutions. By embracing policy changes that prioritize decentralized systems like rooftop solar—through expanded net metering, streamlined permitting processes, and targeted subsidies—governments can foster a cleaner, more resilient energy future. This shift ensures progress without compromising fiscal responsibility or environmental stewardship.

Prioritizing rooftop solar and other decentralized energy sources provides a dual benefit: it supports job creation in emerging sectors while reducing harmful environmental impacts typically associated with large-scale projects. By investing in smaller, community-driven solutions, governments can help create a more balanced, sustainable energy landscape that empowers individuals, cuts reliance on corporate-dominated energy grids, and promotes long-term ecological and economic health. These changes are essential to driving innovation, protecting taxpayers, and safeguarding the environment for future generations.

Chapter 10

Environmental Impact
A Closer Look

As the demand for renewable energy continues to grow, it is crucial to examine the environmental impacts of large-scale projects that are often promoted as the solution to climate change. While these projects aim to protect ecosystems by reducing carbon emissions, they can inadvertently cause significant harm to the environments they occupy. From habitat destruction to carbon offset misconceptions, this chapter explores the environmental costs of large-scale solar and wind farms, comparing them to the more sustainable and localized approach of rooftop solar.

How Large Projects Often Damage the Very Ecosystems They Aim to Protect

Large-scale renewable energy projects like solar farms and wind farms are often promoted as essential solutions for reducing carbon emissions and combating climate change. However, there is an irony in the environmental damage they sometimes inflict on the ecosystems they aim to protect. Solar farms, for instance, require vast tracts of land, often leading to the clearing of habitats that support biodiversity or agricultural activities. In desert ecosystems, such as those impacted by the Ivanpah Solar Project in California, the installation of extensive solar panels has disrupted the delicate habitats of endangered species like the desert tortoise. These tortoises rely on the undisturbed landscape for survival, and the clearing of land and compaction of soil for solar infrastructure poses significant risks to their habitat, mobility, and reproduction. Additionally, the heat generated by the mirrors and solar panels creates "solar flux zones," which can be fatal for birds that fly too close, as well as harmful to other wildlife, potentially altering the entire ecosystem's dynamics (Lovich & Ennen, 2017).

Similarly, wind farms, while crucial for reducing fossil fuel reliance, come with their own set of environmental challenges. The placement of wind turbines has proven particularly harmful to avian and bat populations. Birds, especially raptors and migratory species, are at high risk of fatal collisions with turbine blades, which spin at high speeds. Bats, drawn to turbines for reasons scientists are still investigating, suffer from both collisions and a phenomenon known as barotrauma, where rapid air pressure changes near spinning blades cause fatal injuries. These fatalities can have a significant impact on local bird and bat populations, disrupting the ecological roles they play in pest control, seed dispersal, and pollination. In regions where turbines intersect with migratory bird pathways or habitats of endangered species, these impacts can contribute to population declines that threaten local biodiversity (Pasqualetti, 2019).

The sheer size of these renewable projects often also conflicts with conservation efforts at a local level. To build expansive solar or wind farms, large areas of land must be transformed, undermining habitat protection initiatives and sometimes compromising critical wildlife corridors. The long-term ecological damage includes altered water flow, soil compaction, and changes to vegetation patterns—all of which can take decades, if not longer, to recover. This transformation can weaken local ecosystems' resilience, leaving them more vulnerable to environmental stressors, such as drought and climate fluctuations, which are becoming more common with global climate change.

As the renewable energy sector continues to grow, these large-scale projects raise questions about their alignment with conservation goals. Although they contribute to the reduction of greenhouse gas emissions, the localized environmental impacts they produce highlight the need for more thoughtful, sustainable planning that accounts for ecosystem preservation. Rather than siting projects in sensitive areas, developers could consider less intrusive options, such as rooftop solar, repurposing brownfields, or locating wind farms offshore where they would have a lesser impact on terrestrial wildlife.

The central challenge lies in balancing the undeniable benefits of renewable energy with the urgent need to protect vulnerable ecosystems and species. Ensuring that the transition to clean energy does not inadvertently harm the very environments it seeks to save requires a strategic, conservation-focused approach to renewable energy development. By prioritizing careful site selection, thorough environmental assessments, and the integration of wildlife-friendly technologies, it is possible to drive clean energy goals forward in harmony with nature, rather than at its expense. This balanced approach can lead to a more resilient and sustainable renewable energy landscape—one that truly aligns with both climate and conservation objectives.

Comparisons of the Environmental Footprint of Large Farms vs. Rooftop Solar

When comparing the environmental footprint of large solar and wind farms to rooftop solar, the differences are significant due to the vast land requirements of utility-scale projects. Large solar farms generally require 5 to 10 acres per megawatt of electricity generated, and wind farms need thousands of acres to space turbines properly and optimize wind capture. This extensive land use often leads to habitat loss, soil degradation, and the displacement of land that could be used for agriculture or wildlife conservation (Denholm et al., 2021). In many cases, sensitive ecosystems, such as deserts, are cleared to make room for solar farms, threatening species and disrupting local biodiversity.

In the case of wind farms, the space between turbines often disrupts natural landscapes and wildlife habitats, with additional infrastructure like access roads exacerbating the environmental impact. In agricultural regions, land use for renewable energy projects can displace food production, forcing farmers to either lease or sell their land to energy companies, which reduces agricultural yields and potentially harms the economic well-being of rural communities.

Moreover, large-scale energy projects necessitate substantial transmission infrastructure, often requiring the construction of new power lines to deliver electricity from remote locations to urban centers. This further contributes to environmental degradation as it disrupts more land for transmission corridors, increasing the overall footprint of these projects. Energy transmission losses, due to long-distance transport, also reduce the overall efficiency of these systems.

In contrast, rooftop solar systems bypass many of these challenges. Since they are installed on existing buildings, they don't require additional land clearing or infrastructure. This means that rooftop solar avoids habitat destruction, agricultural displacement, and the need for large-scale transmission networks. The electricity produced by rooftop systems is used close to where it is generated, making it highly efficient while minimizing environmental harm. This contrast highlights rooftop solar as a more sustainable option that maximizes existing space without imposing new ecological costs.

Carbon Offset Myths and Realities in Large-Scale Solar and Wind

Large-scale renewable energy projects are often praised for their carbon-offsetting benefits, but the reality is more nuanced. While solar and wind farms produce clean energy, the processes of construction, installation, and maintenance generate significant carbon emissions. For example, manufacturing photovoltaic (PV) panels requires mining and processing materials like silicon, silver, and aluminum, which are energy-intensive and typically reliant on fossil fuels. Similarly, building wind farms involves substantial carbon footprints from the production of steel, concrete, and rare earth minerals (Hernandez et al., 2015).

The myth of carbon neutrality becomes more complex when considering the lifecycle emissions of large-scale projects. The energy expended in manufacturing, transporting, and installing solar arrays or wind turbines, as well as constructing infrastructure like transmission lines and access roads, must be factored into the carbon balance. While these projects eventually reduce emissions, the initial carbon "debt" incurred during their development can take years, even decades, to offset. This is especially true if the energy used in their construction is derived from non-renewable sources, further complicating the narrative of clean energy.

In contrast, rooftop solar systems have a much smaller carbon footprint during installation and throughout their lifecycle. Since they are installed on existing buildings, they do not require extensive land clearing or large-scale infrastructure. The materials used in rooftop solar installations are also more modest compared to utility-scale projects. Moreover, rooftop solar panels generate electricity close to the point of consumption, avoiding the energy losses associated with transmitting electricity over long distances. This localized production model makes rooftop solar significantly more efficient and environmentally sustainable in the long term, as it minimizes both lifecycle emissions and infrastructure needs

Large-scale renewable projects, while essential in the fight against climate change, often come with significant environmental trade-offs. The destruction of ecosystems, extensive land use, and hidden carbon costs raise critical questions about their overall sustainability. In contrast, decentralized systems like rooftop solar present a far smaller environmental footprint, greater efficiency in land use, and reduced lifecycle emissions.

As the renewable energy sector expands, it is imperative to critically evaluate the scale of these projects against their ecological impacts. By promoting decentralized energy solutions that prioritize both environmental protection and biodiversity, we can create a truly sustainable energy future that benefits both people and the planet. Embracing these changes will not only enhance energy resilience but also safeguard our ecosystems for generations to come.

Chapter 11

Rooftop Solar Success Stories

Rooftop solar has emerged as a transformative force in the renewable energy landscape, demonstrating its potential to empower communities and individuals while driving economic growth. This chapter explores inspiring success stories from various regions that have effectively implemented rooftop solar solutions. By examining the economic benefits for homeowners and local governments, alongside the environmental advantages of decentralized energy systems, we highlight how localized solar initiatives can contribute to a more sustainable and resilient energy future. Through these examples, we will illustrate the profound impact of rooftop solar on both individual lives and the broader community.

Case Studies of Communities and Regions that Have Successfully Implemented Rooftop Solar

Rooftop solar has emerged as a transformative solution across various communities in the United States, proving its effectiveness and potential for widespread adoption. In California, state policies and incentives have played a crucial role in fostering growth in rooftop solar installations, particularly in urban centers. San Diego, for instance, stands out with one of the highest rates of solar adoption in the country, driven by strong local initiatives and a regulatory environment that actively supports clean energy. The city's Solar Incentive Program has empowered countless homeowners to install solar systems, leading to considerable reductions in both energy costs and carbon emissions. As residents shift toward solar energy, they experience increased energy independence and financial savings, which contribute to the region's efforts to reduce reliance on fossil fuels (Hernandez et al., 2015).

In Burlington, Vermont, rooftop solar has been fully embraced as part of the city's ambitious commitment to achieving 100% renewable energy. Burlington's municipal utility has pioneered community solar projects, allowing residents who might otherwise lack access—such as renters and those with shaded roofs—to collectively invest in solar installations. This approach has broadened access to renewable energy, fostering inclusivity and collaboration. By supporting community-based solar initiatives, Burlington has enabled residents to directly contribute to the city's renewable energy targets, while also fostering a sense of shared responsibility and community pride (Solar Energy Industries Association [SEIA], 2023). Burlington's innovative model serves as an example of how municipalities can structure solar projects to increase accessibility, making clean energy viable for a broader range of residents.

These success stories from San Diego and Burlington underscore the importance of community engagement, forward-thinking policies, and local incentives in scaling rooftop solar adoption. As each city harnesses rooftop solar to deliver economic and environmental benefits, they pave the way for a sustainable energy future that is not only achievable but also inclusive. By prioritizing decentralized energy solutions, these communities demonstrate that rooftop solar can thrive in diverse environments, from dense urban settings to small towns. Their experiences highlight the power of local government and grassroots involvement in overcoming traditional barriers to solar adoption, showing that with the right support, rooftop solar can become a cornerstone of the nation's clean energy transformation.

The Economics of Solar: Individual Homeowners and Local Governments Making the Shift

The economics of rooftop solar present a compelling case for both individual homeowners and local governments, offering significant financial and environmental benefits. For homeowners, the initial investment in solar panels is increasingly offset by substantial long-term savings on energy bills. Federal and state incentives, such as tax credits, rebates, and financing programs, help reduce upfront costs, making solar more accessible to a broader range of households. Many homeowners find that, with these incentives, their solar systems can pay for themselves within 5 to 10 years, after which they enjoy virtually free electricity. Additionally, net metering provides an added financial advantage, allowing homeowners to sell any excess energy their systems produce back to the grid. This not only accelerates the payback period but also offers a steady source of income during peak generation periods, enhancing the overall return on investment (National Renewable Energy Laboratory [NREL], 2021).

For local governments, rooftop solar offers a host of economic benefits that align with broader sustainability and economic development goals. As municipalities invest in rooftop solar, they can reduce reliance on fossil fuels and lower municipal energy costs, freeing up funds for other critical services. These investments also contribute to local economic growth, particularly by creating jobs in the installation, maintenance, and servicing of solar systems. Solar energy is a labor-intensive industry, and as communities embrace rooftop solar, they bolster local economies by supporting green jobs and fostering demand for services from local businesses involved in the clean energy sector (Solar Energy Industries Association [SEIA], 2023).

The shift toward solar can also enhance energy resilience, providing a stable, reliable, and distributed energy supply that is less susceptible to disruptions and price volatility than conventional, centralized energy sources. With a more decentralized energy grid, communities become less reliant on distant power plants and high-voltage transmission lines, reducing transmission losses and making the local grid more resilient to extreme weather events and other potential disruptions. Over time, these local investments in rooftop solar yield substantial returns, not only through reduced energy expenses but also by contributing to a robust, energy-independent community.

The combination of reduced energy costs, local job creation, and environmental benefits positions rooftop solar as a win-win solution for homeowners and local governments alike. This model not only advances sustainability goals but also builds economic stability, as communities that embrace rooftop solar are investing in a cleaner, more resilient energy future. With the right incentives and policies, rooftop solar has the potential to become a central component of sustainable urban and rural development, paving the way for a more economically vibrant and environmentally responsible energy landscape.

Why Localized Solutions Provide Better Environmental and Economic Outcomes

Localized energy solutions like rooftop solar provide a multitude of benefits that lead to improved environmental and economic outcomes compared to large-scale renewable projects. First and foremost, rooftop solar minimizes its environmental footprint by utilizing existing infrastructure, thus avoiding the habitat destruction and land-use conflicts often associated with expansive solar farms (Hernandez et al., 2015). This approach not only conserves natural spaces but also helps maintain local biodiversity.

Additionally, decentralized solar generation significantly enhances energy resilience within communities. By producing electricity on-site, households can maintain power during grid outages or emergencies, which is particularly crucial in regions prone to extreme weather events. This localized production reduces strain on centralized power systems, improving overall grid reliability and stability (NREL, 2021).

Economically, localized solar solutions foster greater community engagement and empowerment. When residents invest in rooftop solar, they gain control over their energy production and consumption, allowing them to contribute actively to a more sustainable energy future. This sense of ownership can lead to increased community cohesion as residents collaborate to promote renewable energy initiatives, share knowledge, and support local businesses. Ultimately, the success of rooftop solar not only transforms individual energy practices but also cultivates a collective commitment to sustainability and environmental stewardship

As the success stories of communities and regions adopting rooftop solar continue to multiply, they stand as powerful testament to the potential of localized solutions in driving significant progress toward a sustainable energy future. These initiatives not only demonstrate the feasibility of decentralized energy systems but also showcase their ability to enhance community resilience, reduce environmental impacts, and create economic opportunities. By prioritizing rooftop solar and similar technologies, we can pave the way for a more equitable, sustainable, and economically viable energy landscape that benefits all members of society.

Ultimately, embracing these localized solutions represents a critical step toward achieving energy independence and mitigating climate change. As we learn from these success stories, it becomes clear that the future of energy lies not in large, centralized projects but in empowering individuals and communities to take charge of their energy production. In doing so, we foster a collaborative approach to sustainability that not only protects the planet but also enriches our communities, paving the way for a greener and more resilient tomorrow.

Chapter 12

The Role of Utilities
Adapt or Become Obsolete

The energy landscape in the United States is evolving rapidly, driven by advances in technology, a growing awareness of environmental issues, and shifting consumer preferences. One of the most transformative changes has been the rise of rooftop solar, which has allowed homeowners to generate their own electricity, significantly reducing their reliance on traditional utility companies. As this shift toward decentralized power generation accelerates, utility companies find themselves at a critical juncture: they must either adapt to the new reality or risk becoming obsolete (Darghouth, Wiser, Barbose, & Mills, 2016).

How Utilities Resist Rooftop Solar Adoption to Protect Their Profits

For decades, the business model of utility companies has been based on centralized power generation and distribution, a system that has served as the backbone of energy supply across most of the industrialized world. In this traditional model, large-scale power plants are constructed to generate electricity in bulk. These power plants are often fueled by coal, natural gas, nuclear energy, or large-scale renewable sources like hydroelectric dams and wind farms. Due to the economies of scale, these plants can produce electricity more efficiently and at a lower cost per kilowatt-hour compared to smaller, decentralized generation methods. The electricity is then transmitted over long distances through high-voltage power lines before being distributed to homes, businesses, and industries via local substations and lower-voltage distribution lines.

This centralized structure has allowed utility companies to maintain a monopoly or near-monopoly in many regions, giving them significant control over pricing, distribution, and service. Because the infrastructure required to generate and deliver electricity on such a large scale involves substantial upfront costs, few competitors have been able to enter the market, effectively cementing utilities as the sole providers of electricity in most areas. These utilities are often regulated by state or regional commissions, which approve rate structures and oversee utility operations, but the utilities themselves retain control over the energy supply chain—from generation to transmission to billing.

The traditional model of centralized power generation has several advantages. It enables utilities to manage and maintain large power plants that can generate continuous, reliable power to meet the needs of vast numbers of customers. By centralizing power production, utilities can also plan for and ensure consistent energy availability, balancing supply and demand at the grid level. Moreover, this system has provided a stable revenue stream for utilities, as they charge customers for the electricity they consume, as well as for the upkeep of the grid infrastructure.

However, this centralized model has several drawbacks, including high transmission losses, environmental concerns related to fossil fuel use, and vulnerability to large-scale blackouts when a central node fails. Additionally, the model is inherently inflexible, designed to cater to a one-way flow of electricity from the power plant to the consumer, leaving little room for innovations like decentralized energy generation.

The rise of rooftop solar represents a fundamental challenge to this entrenched model, as it allows individual homeowners and businesses to generate their own electricity, thereby bypassing the centralized grid—at least in part. Solar panels, which convert sunlight into electricity, enable users to produce power right where it is needed, reducing their reliance on the traditional grid. This shift introduces the concept of "prosumers," individuals who both consume and produce electricity, turning the old, one-way model of energy delivery into a dynamic, two-way system. Homeowners with solar panels can not only reduce their dependence on utility-supplied power but also, through net metering, sell excess electricity back to the grid, further undermining the traditional revenue models of utility companies (Borenstein, 2017).

The adoption of rooftop solar challenges utilities in several ways. First, it reduces demand for electricity from the grid, which in turn diminishes the revenue utilities collect through consumption-based pricing. Second, it disrupts the economies of scale that large power plants rely on, as the more homes that generate their own electricity, the less justification there is for continuing to invest in expensive, centralized power generation facilities. Third, it changes the role of the grid itself, from being a central supplier of energy to acting as a backup or balancing mechanism for distributed generation. Finally, it introduces competitive forces into what has historically been a monopoly or heavily regulated industry, forcing utilities to reevaluate their business models to remain financially viable in an increasingly decentralized energy landscape.

In short, rooftop solar fundamentally disrupts the longstanding business model of centralized power generation and distribution, creating both challenges and opportunities for utility companies as they navigate this shift. The success of utilities in this evolving landscape will depend on their ability to adapt to new technologies and embrace more flexible, decentralized energy systems. Those that fail to adapt risk being left behind in an energy future where consumers have more control over their power than ever before.

In response, many utility companies have resisted the widespread adoption of rooftop solar in a variety of ways:

1. **Net Metering Pushback**: Net metering is a policy that allows homeowners with solar panels to sell excess electricity back to the grid, often at retail rates. This not only helps solar users recoup their investment in panels faster but also reduces their monthly energy bills. Utility companies have fought to roll back or limit net metering programs, arguing that solar users are not paying their fair share for the grid's maintenance and infrastructure (Felder & Athawale, 2014). Utilities often claim that the costs of maintaining power lines, transformers, and

other critical infrastructure are spread among all customers, so if fewer people rely on the grid, the burden increases for non-solar customers. However, these arguments are widely seen as attempts to protect profits by stifling competition from rooftop solar (Klein, 2020).

2. **Imposing Additional Fees**: In some regions, utilities have introduced "solar taxes" or additional fees for customers with solar panels. These fees are often justified as necessary to maintain grid reliability, but in practice, they serve to make solar adoption less economically attractive (Hansen, Zakeri, & Syri, 2020). By increasing the cost of connecting to the grid, utilities can dissuade potential rooftop solar customers while ensuring that they continue to collect revenue from those who have already made the switch (Linvill, Shenot, & Lazar, 2013).

3. **Lobbying and Legislative Influence**: Utility companies have deep pockets and often wield significant political influence at the state and national levels. They have used this influence to lobby for laws and regulations that hinder the growth of rooftop solar. For example, some utilities have pushed for caps on the number of homes that can participate in net metering programs, or they have supported legislation that requires homeowners to navigate complex regulatory hurdles before installing solar panels (Bird, Reger, Heeter, & O'Shaughnessy, 2018). These tactics slow the adoption of rooftop solar and preserve the utilities' dominance over power generation and distribution.

The Tension Between Centralized Power Generation and Home-Based Solar

The conflict between utility companies and the solar industry represents a deeper tension between two fundamentally different approaches to power generation: centralized versus decentralized energy. This divide is not merely about technology; it reflects a profound shift in how energy is produced, distributed, and consumed. At its core, the tension speaks to a struggle over control, cost structures, infrastructure investment, and, ultimately, the future of the energy grid.

1. **Centralized Power Generation**: Historically, energy production has been centralized in large power plants that can generate massive amounts of electricity at once. These plants benefit from economies of scale and can produce electricity more cheaply than smaller, decentralized systems (Stirling, 2014). However, centralized generation also has its drawbacks, including vulnerability to large-scale blackouts, high transmission costs, and significant environmental impacts, especially when fossil fuels are the primary energy source (Fell, 2019).

2. **Home-Based Solar (Decentralized Energy)**: Rooftop solar represents the opposite model: small-scale, localized energy production. Solar panels generate electricity on-site, reducing the need for long-distance transmission and minimizing energy losses (Seel, Barbose, & Wiser, 2014). Home-based solar systems also offer resilience in the face of grid outages, as homeowners with battery storage systems can keep the lights on even when the central grid goes down (Baker, 2019). Moreover, rooftop solar is inherently greener, as it eliminates the need for large fossil fuel-burning power plants and reduces greenhouse gas emissions (Millstein, Wiser, Bolinger, & Barbose, 2017).

This tension between centralized and decentralized energy is not just a technical or economic issue—it is also a question of control. Utilities have traditionally controlled the entire energy supply chain, from generation to distribution. Rooftop solar empowers individual homeowners to generate their own energy, eroding the utilities' control and forcing them to compete in a more open and dynamic marketplace (Joskow, 2019).

Potential Paths Forward for Utility Companies in a Decentralized Energy Landscape

Despite their resistance, utilities cannot ignore the rise of rooftop solar and other forms of decentralized energy forever. As more homeowners adopt solar, the traditional utility model will become less and less viable. To survive in this new landscape, utilities must adapt. Several potential paths forward could allow them to thrive in a decentralized energy future:

1. **Embracing the Role of Grid Managers**: In a decentralized energy system, the grid becomes even more important. Instead of generating most of the electricity, utilities could shift their focus to managing the flow of electricity between homes, businesses, and other power producers (Owens & Faruqui, 2011). This would require investing in "smart grid" technology that can handle two-way energy flows, as well as integrating distributed energy resources like solar panels, batteries, and electric vehicles. By providing reliable grid services, utilities could maintain a steady revenue stream even as their role in power generation diminishes (Sioshansi, 2020).

2. **Investing in Renewable Energy**: Utilities can stay competitive by embracing renewable energy themselves. Large-scale solar farms, wind farms, and other renewable projects still offer significant opportunities for centralized power generation. By investing in these projects, utilities can continue to provide clean energy to customers who are not able or willing to install rooftop solar. This approach would allow utilities to remain relevant in a carbon-constrained world while meeting the growing demand for renewable energy (Lazard, 2019).

3. **Offering Solar and Storage Solutions**: Some forward-thinking utilities have begun offering their customers solar panel installations, battery storage systems, and energy management services. By entering the rooftop solar market themselves, utilities can turn a potential threat into a new revenue stream. Additionally, by providing storage solutions, utilities can help customers maximize the value of their solar

panels, ensuring that excess energy is stored for use during cloudy days or at night (Smith, 2019). Utilities that embrace this approach will position themselves as leaders in the transition to a more distributed energy future.

4. **Developing New Pricing Models**: The traditional utility pricing model, based on kilowatt-hour consumption, may not work in a future where many customers generate their own electricity. To adapt, utilities will need to develop new pricing models that reflect the changing energy landscape. For example, some utilities are exploring time-of-use pricing, where customers are charged different rates depending on when they use electricity. This encourages customers to use energy during off-peak times, which helps balance supply and demand on the grid (Borenstein, 2005). Other utilities are experimenting with subscription-based models, where customers pay a flat fee for access to the grid, regardless of how much electricity they use. By developing more flexible pricing models, utilities can remain profitable while supporting the growth of decentralized energy (Hogan, 2014).

Evolve or Become Obsolete Conclusion

The rise of rooftop solar and decentralized energy marks a pivotal turning point for the energy industry, representing not just a technological shift but a profound rethinking of how energy is generated, distributed, and consumed. Traditional, centralized business models—where large-scale power plants supply energy through one-way electricity flows—are becoming increasingly outdated in the face of a rapidly evolving energy landscape. With individuals, communities, and businesses now able to generate their own power through rooftop solar, battery storage, and microgrids, utility companies can no longer assume their historical role as the sole providers of electricity. Instead, they face a future where decentralized energy systems take center stage, fundamentally reshaping how energy is managed and supplied.

The need for adaptation is clear. Utility companies must embrace new business models to remain relevant in a world where the grid is no longer the only source of power. This shift calls for utilities to invest in renewable energy sources such as large-scale solar and wind, while also reimagining their role as facilitators of a more distributed energy network. By integrating decentralized technologies like rooftop solar and battery storage into their strategies, utilities can support a grid that is increasingly interactive and consumer-driven. New revenue models—such as offering energy management services, developing microgrids, or supporting community solar initiatives—allow utilities to remain central to the energy ecosystem, but in ways that empower rather than monopolize. Such approaches require flexibility and a customer-centric mindset, recognizing that today's energy consumers are more empowered and expect more options than ever before.

However, if utilities resist this transformation, they risk becoming obsolete in an energy revolution that is already well underway. The forces driving this shift—falling costs of solar panels, advancements in battery technology, rising environmental consciousness, and supportive government policies—are too significant to ignore. Utilities that fail to innovate face losing customers to alternative energy sources, as more individuals and businesses choose to generate their own electricity and reduce their dependence on traditional providers. Revenue streams may shrink, regulatory pressures to decarbonize will intensify, and consumers increasingly value sustainability, resilience, and independence. Without innovation, these companies may find themselves sidelined in an industry where consumers and communities seek control and sustainability.

In this evolving energy paradigm, clinging to outdated models will not only hinder utilities' relevance but may also exacerbate issues decentralized energy seeks to resolve, such as energy inequality, environmental harm, and grid vulnerability. Conversely, utilities that innovate, adapt, and take a leadership role in the decentralized energy revolution will actively shape the future of the industry. These forward-thinking companies will be instrumental in developing the smart grids of tomorrow, empowering consumers to take control of their energy needs, and building more resilient systems that can handle the complexities of the 21st century, from climate change to fluctuating energy demands.

The choices utilities make now will determine whether they thrive or become cautionary tales in the story of energy transformation. The path is clear: adapt, evolve, and lead the way toward a decentralized, resilient energy future, or risk obsolescence as the energy transition accelerates. Those who embrace this shift will emerge as architects of a cleaner, more dynamic energy economy, positioning themselves as indispensable players in a decentralized energy landscape. Those who resist, however, will be left behind, watching as their traditional models and infrastructure become relics in the dust of an accelerating energy transition.

Chapter 13

Looking Ahead
A New Energy Model for America

The energy landscape in the United States is on the brink of a profound transformation. With the rise of rooftop solar and decentralized energy systems, the possibility of a future where homeowners and businesses generate their own electricity is no longer a distant dream—it is an emerging reality. The traditional, centralized energy model that has dominated for over a century is being disrupted by advances in renewable technology, declining costs, and the growing recognition of the need for a sustainable, resilient energy grid. As America looks ahead, the question is not whether decentralized energy will play a role but how quickly and effectively it can be integrated into the mainstream energy system.

Envisioning an America Where Rooftop Solar and Decentralized Energy Lead the Way

In a future where rooftop solar and decentralized energy lead the way, the energy grid would undergo a radical transformation, fundamentally changing from the traditional top-down, centralized infrastructure to a more fluid, interconnected, and locally driven network. Today's energy system relies on a relatively small number of large power plants that produce electricity centrally and distribute it over long distances to homes and businesses. These centralized power plants, often powered by fossil fuels or nuclear energy, operate on a large scale, generating immense amounts of electricity that are fed into the grid and transmitted to consumers through an extensive network of high-voltage transmission lines and local distribution systems.

However, in a decentralized energy future, this model would be replaced by a far more diverse and dispersed energy generation system. Instead of depending on a limited number of massive power plants, millions of small, distributed energy sources would supply the country's electricity needs. Homeowners would install rooftop solar panels to power their homes, and businesses would adopt solar or wind technologies to meet their energy demands. In rural areas, small-scale wind turbines could be installed to take advantage of local wind resources, while community solar farms would allow those without suitable rooftops to participate in clean energy production. This distributed energy model would include not only solar panels and wind turbines but also other renewable energy resources like geothermal systems, small hydropower projects, and energy storage technologies such as batteries and fuel cells (Borenstein, 2017).

These systems would be interconnected through smart grid technologies—digital systems that allow for real-time monitoring, communication, and control of energy flows. Smart grids would be essential for managing the complex interactions between millions of decentralized energy resources (DERs) and the grid. They would enable utilities and grid operators to balance supply and demand more efficiently, automatically adjusting the flow of electricity based on real-time consumption patterns, weather conditions, and energy generation from distributed sources. Smart meters, which monitor household energy use in real time, would provide critical data to both consumers and utilities, allowing for better energy management and pricing strategies (Sioshansi, 2020). Battery storage systems would also play a crucial role, enabling households and businesses to store excess energy generated by their solar panels or other DERs for use during periods when the sun isn't shining or when energy demand is high. These storage systems would help stabilize the grid by providing backup power and reducing the need for peaker plants—high-cost, high-emission power plants that are typically used to meet peak electricity demand (Millstein et al., 2017).

In this envisioned future, the flow of energy would no longer be unidirectional, where utilities generate electricity and consumers passively receive it. Instead, the energy flows would become dynamic and multidirectional. Consumers would become "prosumers," actively contributing to the energy system by generating their own electricity, storing it, and even selling surplus energy back to the grid. In many cases, energy would not just flow from large power plants to consumers but also from home solar systems to the grid, from community microgrids to local businesses, and even between neighbors through peer-to-peer energy trading platforms (Owens & Faruqui, 2011). This would create a decentralized web of energy flows, allowing communities to be more energy-independent and less reliant on distant power plants.

Localized microgrids would further enhance energy security and resilience by enabling neighborhoods, towns, or even entire regions to operate independently from the national grid during emergencies or outages. Microgrids are self-contained energy systems that can either operate in conjunction with the broader grid or disconnect and function autonomously in "island mode" when necessary. In the event of a natural disaster, cyberattack, or grid failure, microgrids would ensure that critical services like hospitals, emergency response centers, and water treatment facilities remain powered. This decentralized, community-based energy approach would also allow households to maintain access to electricity during extreme weather events, increasing resilience against the growing threat of climate change (Baker, 2019).

This new energy model would fundamentally alter the relationship between consumers and utilities. Utilities, which today operate primarily as suppliers of electricity, would evolve into facilitators and managers of energy transactions. Rather than competing with rooftop solar and decentralized energy systems, forward-thinking utilities would embrace this shift by providing new services to help consumers manage their energy production and consumption. Utilities could offer energy management services, such as helping homeowners install solar panels, optimizing energy storage, or advising businesses on how to maximize their energy efficiency. They could also act as brokers of decentralized energy, buying excess electricity from prosumers and selling it to others in need, thus creating new revenue streams and fostering greater collaboration between energy producers and consumers (Sioshansi, 2020).

Moreover, utilities would play a critical role in managing the integration of decentralized energy resources into the broader grid. As the number of distributed energy sources increases, utilities would be responsible for ensuring that the grid remains stable and reliable, despite the fluctuating nature of renewable energy generation. This would involve managing two-way energy flows, balancing intermittent energy production from solar and wind with energy storage systems, and optimizing demand response programs that incentivize consumers to adjust their energy use during peak times (Lazar & Gonzalez, 2015). By developing these new capabilities and embracing their role as energy managers, utilities would remain relevant and indispensable in the decentralized energy future.

The result of this shift would be a democratized energy system, where individuals and communities have far greater control over their energy use and production. The traditional utility-customer relationship, characterized by passive consumption and centralized control, would give way to a more collaborative, participatory energy ecosystem. Individuals, empowered by renewable energy technologies and smart grids, would become active participants in managing the energy they use and produce. This would not only reduce reliance on fossil fuels but also minimize the environmental impacts of energy generation. With less need for large, centralized power plants and fewer transmission losses, the carbon footprint of electricity production would shrink, contributing to significant reductions in greenhouse gas emissions and helping to mitigate climate change (Millstein et al., 2017).

In conclusion, a future driven by rooftop solar and decentralized energy would represent a profound departure from the energy systems of the past. The energy grid would become more localized, flexible, and resilient, with millions of small energy producers contributing to a dynamic and interconnected network. Utilities, once dominant suppliers of electricity, would evolve into facilitators of energy transactions, helping to manage and integrate decentralized energy resources into the grid. This future holds the promise of greater energy independence for individuals and communities, reduced environmental impacts, and a more resilient and equitable energy system for all.

Policy Changes Needed to Encourage Widespread Adoption of Rooftop Solar

To make this vision a reality, significant policy changes will be required at both the federal and state levels. The U.S. government must create a regulatory environment that encourages the widespread adoption of rooftop solar and other decentralized energy solutions, addressing the barriers that currently hinder their growth.

1. **Strengthening Net Metering Programs**: One of the most effective ways to incentivize rooftop solar adoption is through net metering, which allows homeowners with solar panels to sell excess electricity back to the grid. However, many states have begun scaling back their net metering programs, reducing the economic benefits of installing solar panels. Policymakers must ensure that net metering remains robust and that utilities fairly compensate solar users for the energy they contribute to the grid (Darghouth, Wiser, Barbose, & Mills, 2016). Additionally, states should establish clear and transparent guidelines to prevent utilities from imposing unfair fees on solar users, ensuring that rooftop solar remains a financially viable option for all Americans.

2. **Tax Incentives and Subsidies**: Federal and state tax incentives have played a critical role in accelerating solar adoption, and these incentives must continue to be supported and expanded. The federal Investment Tax Credit (ITC), which provides a tax credit for a percentage of the cost of installing solar energy systems, has been instrumental in reducing upfront costs for consumers. As the ITC phases out, it will be essential for policymakers to either extend the credit or implement alternative incentives that encourage continued investment in rooftop solar (Borenstein, 2017). In addition, state governments should offer property tax exemptions, rebates, and other subsidies to make rooftop solar accessible to a broader range of consumers, particularly low- and middle-income households.

3. **Modernizing the Grid and Regulatory Framework**: The current regulatory framework for utilities was designed for a centralized energy model, and it must be updated to reflect the realities of a decentralized, renewable-based energy system. Policymakers should encourage investments in smart grid technology, which will allow utilities to better manage two-way energy flows, integrate DERs, and improve grid reliability. Furthermore, regulators must work with utilities to develop new pricing models that reflect the changing nature of energy consumption. Time-of-use pricing, for example, could encourage consumers to use electricity during off-peak hours, helping to balance supply and demand more effectively (Hogan, 2014).

4. **Expanding Access to Community Solar Programs**: Not every American has the ability to install solar panels on their home, whether due to financial constraints, shading issues, or renting instead of owning their home. To address this, policymakers should promote the development of community solar programs, which allow multiple participants to share the benefits of a single solar array. These programs provide a crucial opportunity for individuals who may not have access to rooftop solar to still participate in the clean energy transition and reduce their electricity bills (Millstein, Wiser, Bolinger, & Barbose, 2017).

How This Transition Could Benefit the Economy, Environment, and Future Generations

The transition to a decentralized energy system centered around rooftop solar would deliver significant benefits for the economy, the environment, and future generations.

1. **Economic Growth and Job Creation**: Decentralized energy systems would spur innovation and create jobs in the growing renewable energy sector. The solar industry alone has already become one of the fastest-growing sectors in the U.S. economy, employing hundreds of thousands of workers across installation, manufacturing, research, and sales. As rooftop

solar continues to expand, the demand for skilled labor will rise, creating new job opportunities for electricians, engineers, project managers, and solar panel manufacturers (Seel, Barbose, & Wiser, 2014). Moreover, the proliferation of decentralized energy could drive economic growth by reducing energy costs for consumers and businesses, freeing up capital for other investments.

2. **Environmental Benefits**: One of the most compelling reasons to promote rooftop solar is its potential to drastically reduce greenhouse gas emissions. By generating electricity from the sun, rooftop solar eliminates the need to burn fossil fuels, reducing the amount of carbon dioxide and other harmful pollutants released into the atmosphere. A widespread transition to decentralized, renewable energy would play a pivotal role in combating climate change, improving air quality, and protecting public health (Millstein et al., 2017). The environmental benefits of this transition would be felt not just in the U.S. but around the world, as global efforts to decarbonize the energy sector accelerate.

3. **Resilience and Energy Security**: A decentralized energy system offers greater resilience against natural disasters, cyberattacks, and other threats to the traditional grid. In the face of increasingly frequent and severe weather events driven by climate change, localized energy production through rooftop solar and microgrids would provide a critical safeguard against widespread blackouts. Battery storage systems would enable homes and businesses to store excess energy for use during outages, ensuring continuity of power even when the central grid is compromised. This enhanced resilience would be particularly valuable for vulnerable communities and critical infrastructure, such as hospitals, emergency services, and food supply chains (Baker, 2019).

4. **A Legacy for Future Generations**: Perhaps the most important benefit of transitioning to a decentralized energy system is the legacy it would leave for future generations. By

investing in renewable energy today, the U.S. can ensure a cleaner, healthier, and more sustainable planet for tomorrow's children. Future generations would inherit a more resilient and flexible energy grid, capable of meeting the challenges posed by climate change, population growth, and evolving energy needs. Additionally, they would benefit from the long-term economic and environmental stability that comes from reducing dependence on finite, polluting fossil fuels (Joskow, 2019).

Looking ahead, it is clear that America's energy future lies in decentralized systems like rooftop solar. These systems represent more than just a technological shift; they symbolize a transformation in how energy is produced, consumed, and managed. Unlike the centralized energy systems of the past, which rely on large power plants and long-distance transmission lines, decentralized systems empower individuals, communities, and businesses to generate their own electricity closer to where it is consumed. This shift not only increases efficiency and reduces energy loss but also fosters energy independence by allowing consumers to take control of their energy needs, helping to democratize the energy landscape.

By enacting the necessary policy changes and embracing a more distributed energy model, the United States can set the stage for a cleaner, more resilient, and more equitable energy system. Rooftop solar, in particular, offers the potential to decentralize power generation, reduce greenhouse gas emissions, and improve local energy resilience by reducing dependence on centralized grids. In regions prone to natural disasters or grid vulnerabilities, decentralized systems like rooftop solar, coupled with battery storage, can provide critical backup power during outages, enhancing overall grid security. Additionally, this energy model allows consumers, including low- and middle-income households, to directly benefit from clean energy through financial savings and reduced utility bills, fostering greater economic equity.

The environmental benefits of embracing decentralized systems are clear. By prioritizing renewable energy sources like rooftop solar, the U.S. can significantly reduce its carbon footprint, contributing to global efforts to mitigate climate change. This transition would also reduce harmful air pollution from fossil fuel plants, improving public health outcomes, particularly in disadvantaged communities that are often disproportionately affected by pollution. Furthermore, decentralizing energy production reduces the need for expansive land use and the environmental degradation associated with large-scale power plants and transmission infrastructure.

However, this transition to a decentralized energy future requires coordinated effort from all key stakeholders: utility companies, policymakers, and consumers. Utility companies, which have historically relied on centralized energy generation, must recognize that their business models need to evolve. Rather than viewing decentralized energy systems as competition, they can play a pivotal role in facilitating this transition by investing in grid modernization, energy storage solutions, and smart grid technologies that enable seamless integration of distributed energy resources. Additionally, utilities can offer value-added services, such as energy management and consulting for consumers looking to adopt rooftop solar, creating new revenue streams in this evolving landscape.

Policymakers must take a proactive role in removing regulatory barriers that hinder the growth of decentralized energy. This includes expanding access to net metering programs, offering tax incentives and subsidies for rooftop solar installations, and ensuring that utility regulations support the development of microgrids and distributed energy resources. Policymakers can also address equity concerns by ensuring that low-income communities have access to decentralized energy technologies through programs like community solar and targeted financial assistance.

Consumers have a crucial role to play as well. By adopting rooftop solar and other decentralized energy solutions, individuals can reduce their carbon footprints and contribute to a cleaner, more sustainable energy grid. Public awareness campaigns and education efforts will be essential to encourage more widespread adoption, particularly as more people realize the long-term financial and environmental benefits of these systems.

The path to a sustainable energy future is clear—now it is up to us to take the necessary steps to get there. This requires bold leadership, innovative policy frameworks, and widespread collaboration among all sectors of society. The transition to decentralized energy systems is not just a technological imperative but a moral one, as it offers a unique opportunity to create a more just and sustainable energy system that benefits not only the present generation but also future generations to come.

Chapter 14

Rooftop Solar

The Most Effective Path to Sustainable Energy

In the pursuit of a sustainable energy future, rooftop solar has emerged as a powerful solution that balances energy needs with environmental responsibility. Unlike traditional large-scale energy projects, which often require extensive land and resources, rooftop solar harnesses the sun's energy directly at the point of use, avoiding many of the environmental and logistical pitfalls associated with centralized power generation. This chapter explores why rooftop solar is the most effective mode of alternative energy, its unique benefits, and how its minimal impact on the natural environment positions it as an optimal choice for the United States and beyond.

Decentralized Power Generation: Efficient and Resilient

Rooftop solar operates on a decentralized model, generating electricity directly on individual homes, businesses, and public buildings rather than at a distant central power station. This approach minimizes energy loss during transmission and reduces dependency on long, complex grid networks, making rooftop solar inherently more efficient. In traditional centralized power systems, electricity generated at large-scale power plants must travel long distances through transmission lines before reaching end users. This journey requires extensive infrastructure investments and incurs significant energy losses due to resistance in transmission lines, with an estimated 5-10% of energy lost on average during transmission (National Renewable Energy Laboratory [NREL], 2021). By producing energy directly at the point of consumption, rooftop solar not only maximizes the efficiency of each kilowatt-hour generated but also lowers costs associated with transmission infrastructure. For homeowners and businesses, this translates to cost savings on utility bills, as more of the generated energy is retained and consumed locally (Carley & Konisky, 2020).

Additionally, the decentralized nature of rooftop solar enhances community resilience, providing a critical advantage in the face of growing grid vulnerabilities and increasing extreme weather events. When natural disasters or technical failures disrupt centralized power systems, communities with rooftop solar installations can maintain partial or even full power, especially when coupled with battery storage systems (EnergySage, 2021). Batteries allow excess energy generated during peak sunlight hours to be stored for use at night or during outages, giving homes and businesses a buffer against power interruptions. In regions prone to severe weather, such as hurricanes, wildfires, or snowstorms, this resilience is invaluable, as centralized grids are often susceptible to widespread outages during such events. Rooftop solar systems enable communities to remain functional and safer during prolonged power outages, supporting emergency response and providing power for critical needs (Hernandez et al., 2015).

Rooftop solar contributes not only to individual energy independence but also to broader community security and stability. With distributed generation, communities are less dependent on large utilities and vulnerable grid systems, empowering residents to take control of their energy needs and reducing the risk of large-scale disruptions. This decentralized model aligns with trends toward microgrids—localized grids that can operate independently of the central grid—which further enhances resilience and allows communities to optimize their energy use based on real-time data and localized demand (Solar Energy Industries Association [SEIA], 2023).

As rooftop solar expands, it also paves the way for the integration of additional sustainable technologies. For example, electric vehicle (EV) owners can charge their cars with solar-generated electricity, further reducing reliance on fossil fuels. Additionally, rooftop solar systems can be combined with smart home technology, allowing homeowners to monitor and optimize their energy use, enhancing both efficiency and convenience (Lovich & Ennen, 2017). These technologies create a self-sustaining energy ecosystem within individual households and communities, contributing to reduced greenhouse gas emissions and a cleaner environment overall.

In this sense, rooftop solar serves as a cornerstone of energy security and community stability, transforming homes and businesses into mini power stations capable of supporting themselves and even their neighbors during times of need. The transition to decentralized energy empowers individuals, fosters community resilience, and shifts society closer to a future in which energy generation is clean, reliable, and locally controlled.

Limited Land Impact: Conserving Natural Habitats and Resources

One of the most significant environmental advantages of rooftop solar is its minimal land footprint. Large-scale solar farms and wind projects often require vast tracts of land, which can displace habitats and alter ecosystems in the process. For example, solar installations in sensitive environments like deserts and grasslands can disrupt fragile ecosystems, impacting species such as the endangered desert tortoise, which depends on undisturbed habitats for survival (Lovich & Ennen, 2017). Similarly, wind farms, while beneficial for reducing carbon emissions, pose risks to migratory birds and bats, which are susceptible to fatal collisions with turbine blades. Furthermore, these installations require extensive open spaces, often placing them in areas that overlap with critical wildlife corridors and natural landscapes, which may take years to recover.

In contrast, rooftop solar makes efficient use of already-developed spaces—such as rooftops, parking structures, and public buildings—making it particularly suitable for urban and suburban settings. By repurposing these existing surfaces for energy production, rooftop solar avoids the need to clear natural landscapes, effectively preserving biodiversity and protecting wildlife habitats. This model of energy generation respects the integrity of natural spaces and aligns with sustainable land use principles, reducing the ecological impact typically associated with traditional energy infrastructure. Instead of requiring new land and disrupting ecosystems, rooftop solar leverages human-made structures, transforming them into valuable sources of renewable energy with minimal environmental cost. This approach not only conserves natural habitats but also aligns with urban planning initiatives aimed at maximizing the utility of developed spaces while reducing cities' carbon footprints.

By focusing on rooftop solar as a primary mode of alternative energy, communities can support renewable energy goals without sacrificing critical habitats or contributing to habitat fragmentation. Rooftop solar's minimal land footprint makes it a more ecologically responsible choice for densely populated areas and regions with delicate ecosystems, helping to maintain biodiversity and reduce the negative environmental impacts commonly associated with large-scale renewable energy installations.

Reduced Environmental Degradation: Protecting Air, Water, and Soil

Beyond land conservation, rooftop solar also reduces the environmental degradation commonly associated with traditional energy production. Unlike fossil fuels, which produce harmful emissions and contribute to air pollution, rooftop solar operates quietly and cleanly, producing no greenhouse gases during its operation (National Renewable Energy Laboratory [NREL], 2021). Fossil fuel-based power plants release carbon dioxide, sulfur dioxide, nitrogen oxides, and particulate matter into the atmosphere, which not only contribute to climate change but also degrade air quality, posing health risks to nearby communities (U.S. Environmental Protection Agency [EPA], 2022). In contrast, rooftop solar systems generate electricity directly from sunlight without combustion, resulting in a zero-emissions energy source.

Rooftop solar also avoids several of the negative environmental impacts associated with large-scale renewable projects. Solar farms, particularly those located in desert ecosystems, often require substantial land clearing and infrastructure, which can lead to soil compaction and erosion. Soil compaction from heavy equipment disrupts the soil's natural structure, reducing its capacity to absorb water and impacting local hydrology. This disturbance can alter water cycles, affecting nearby vegetation and reducing habitats for local wildlife, including endangered species like the desert tortoise (Lovich & Ennen, 2017). Over time, these changes can lead to long-term ecological impacts that degrade the surrounding environment, including increased erosion, loss of native plant species, and habitat fragmentation.

Rooftop solar, by leveraging existing structures such as rooftops, parking structures, and other built environments, avoids these issues entirely. Once installed, rooftop solar systems do not require further land, water, or resource-intensive processes, making them a truly low-impact energy option. Unlike some traditional energy sources and even large-scale renewables, rooftop solar does not consume water resources for cooling or other operational needs, which is especially valuable in water-scarce regions (NREL, 2021). This conservation of water aligns rooftop solar with broader environmental goals, as it reduces strain on local ecosystems and maintains water availability for natural habitats and agriculture.

By generating electricity without impacting water resources or soil health, rooftop solar contributes to cleaner air, soil, and ecosystems, aligning with both conservation efforts and climate goals. Its minimal resource use and negligible operational impact make it one of the most environmentally friendly energy sources available, providing communities with a sustainable way to meet their energy needs while protecting local ecosystems from degradation.

Energy Independence and Local Economic Benefits

Rooftop solar fosters energy independence, enabling homeowners, businesses, and even municipalities to control their own energy production. By installing solar panels, individuals can generate their own electricity, reducing reliance on centralized power plants and creating a financial buffer against rising energy costs. This empowerment fosters a sense of ownership over energy resources, aligning with the growing consumer demand for sustainable, self-sufficient solutions. In a world where utility rates continue to fluctuate, rooftop solar offers individuals and communities greater control over their energy expenses, making it a practical and financially advantageous choice for households and local governments alike (Solar Energy Industries Association [SEIA], 2023).

Additionally, rooftop solar provides substantial economic benefits, as local installation and maintenance jobs help boost local economies. According to the National Renewable Energy Laboratory (NREL), every megawatt of installed rooftop solar creates numerous jobs in the fields of installation, operations, maintenance, and technical support, directly contributing to job growth within the renewable energy sector (NREL, 2021). Federal and state incentives, such as tax credits and rebates, make rooftop solar increasingly accessible to households across income levels, further fueling adoption. Programs like net metering, which allow homeowners to sell any excess energy they produce back to the grid, provide financial returns and help offset the initial costs of installation over time (Carley & Konisky, 2020). These incentives make rooftop solar a more viable option for households that may have previously found clean energy financially out of reach.

This economic accessibility and the local investment model make rooftop solar an equitable renewable energy solution, as it supports a broader transition to sustainable energy for diverse populations. The inclusivity of rooftop solar—enabled by rebates, net metering, and declining installation costs—means that clean energy is increasingly attainable for a wide range of households, including lower-income families. Community solar projects, which allow renters and individuals without suitable rooftops to invest in shared solar installations, further expand access to renewable energy. By supporting inclusive access, rooftop solar promotes environmental justice by providing clean, affordable energy for communities traditionally underserved by large-scale energy projects.

Scalability and Future Potential: Adapting to Urban and Rural Environments

The scalability of rooftop solar makes it uniquely adaptable across a wide variety of settings, from dense urban neighborhoods to remote rural communities. Unlike large-scale renewable energy projects, which often struggle to find suitable land near high-demand areas due to space constraints or environmental considerations, rooftop solar can be installed on existing buildings wherever they are located. This capability allows rooftop solar to seamlessly fit into metropolitan landscapes, suburban zones, and rural villages alike, meeting the energy needs of diverse communities and providing clean energy solutions to areas that may otherwise lack access to renewable power (National Renewable Energy Laboratory [NREL], 2021). This versatility makes rooftop solar a truly inclusive form of renewable energy, with the potential to reduce energy inequalities by bringing sustainable power to underserved and remote areas without the need for extensive infrastructure.

As technology advances, rooftop solar systems are becoming increasingly efficient and cost-effective. Innovations in solar panel design, battery storage, and energy management systems are significantly enhancing the performance and longevity of rooftop installations. For example, advancements in photovoltaic technology have improved solar panels' ability to capture sunlight in lower-light conditions, expanding their usability across different climates and geographical locations. Modern battery storage solutions, meanwhile, allow households and businesses to store surplus energy for use during the night or during peak-demand periods, maximizing the benefits of each kilowatt-hour generated. These improvements make rooftop solar a practical choice for a broader array of homes and businesses, enabling users to maximize their energy investment while simultaneously reducing their environmental footprint (Solar Energy Industries Association [SEIA], 2023).

The continued reduction in installation and maintenance costs, driven by both technological advancements and growing demand, positions rooftop solar as one of the most cost-effective solutions in the renewable energy sector. The energy output of rooftop solar has steadily increased over recent years, meaning today's panels can generate significantly more power than older models on the same footprint (EnergySage, 2022). This combination of increased efficiency and lower costs not only supports wider adoption but also enhances the financial viability of rooftop solar for households across income levels, as lower installation costs and improved efficiency translate to quicker payback periods and greater long-term savings.

These technological and economic advancements make rooftop solar a leading solution for sustainable energy in the renewable sector. Its flexibility and scalability make it ideal for urban and rural settings alike, while ongoing improvements ensure it remains a competitive, viable choice for households and communities aiming to reduce their carbon footprint. By fostering clean energy adoption across diverse settings, rooftop solar supports a more inclusive and sustainable energy transition that benefits both people and the planet.

Policy Support and Community Engagement: The Path Forward

For rooftop solar to reach its full potential, supportive policies and active community engagement are crucial. Local, state, and federal governments have the power to facilitate rooftop solar adoption through various incentives, streamlined permitting processes, and net metering programs that allow homeowners to earn compensation for excess energy they produce. For example, tax credits and rebates reduce the upfront cost of installation, making solar systems more accessible to a wider range of households. Additionally, net metering policies enable homeowners to "sell" surplus energy back to the grid, further offsetting costs and making rooftop solar a more financially viable option (Solar Energy Industries Association [SEIA], 2023). By establishing policies that prioritize rooftop solar over expansive, land-intensive energy projects, governments can help mitigate the environmental degradation that large-scale installations often bring, while supporting a shift toward more sustainable and decentralized energy practices.

Community engagement plays an equally vital role in expanding rooftop solar adoption. Through public education on the benefits of decentralized energy, such as lower utility bills, enhanced energy independence, and reduced carbon footprints, communities can build a culture that embraces sustainability. Educational programs and community forums can inform residents about the environmental and economic advantages of rooftop solar, helping to dispel misconceptions and generate broader interest. Furthermore, initiatives that provide financial support to low-income households—such as grants, rebates, or low-interest loans—ensure that rooftop solar becomes accessible to a broader demographic, democratizing energy generation and helping reduce energy inequality (Carley & Konisky, 2020).

Programs that encourage the development of community solar projects further expand access, especially for renters and individuals whose rooftops may not be suitable for solar installation. Community solar initiatives allow residents to invest in or subscribe to a shared solar installation, benefiting from renewable energy even if they cannot install it on their own property. These initiatives are essential for making rooftop solar a realistic option for a diverse range of households, enhancing inclusivity within the clean energy movement and fostering environmental stewardship (National Renewable Energy Laboratory [NREL], 2021).

In sum, for rooftop solar to fulfill its potential as a transformative energy source, policies that provide financial incentives, simplify permitting, and encourage net metering are essential. When combined with robust community engagement, these policies can accelerate the transition to clean energy, promote energy equity, and support the growth of decentralized energy solutions that empower communities and protect the environment.

Rooftop Solar as the Future of Sustainable Energy

In the pursuit of a cleaner, more sustainable energy future, rooftop solar emerges as a transformative solution that offers an efficient, environmentally friendly alternative to expansive renewable projects. By harnessing the sun's energy directly at the point of consumption, rooftop solar maximizes every watt of power produced, reduces transmission losses, and minimizes the infrastructure burden on our landscapes. Unlike large-scale solar or wind farms that often require vast tracts of land and disrupt natural habitats, rooftop solar leverages existing structures—homes, businesses, schools, and public buildings—preserving open spaces, protecting biodiversity, and empowering communities to take control of their energy needs. This unique combination of efficiency, low environmental impact, and community resilience positions rooftop solar as an unparalleled driver in America's renewable energy transformation (National Renewable Energy Laboratory [NREL], 2021).

Beyond just a technological advancement, rooftop solar represents a shift toward a decentralized, community-centered energy model that values environmental integrity, local empowerment, and economic inclusivity. By generating power locally, rooftop solar allows homeowners and communities to reduce dependency on centralized utilities, creating energy independence that strengthens neighborhoods and builds resilience against grid vulnerabilities and rising energy costs (Carley & Konisky, 2020). This shift is particularly vital in areas prone to extreme weather, where centralized power grids are often vulnerable; rooftop solar enables homes and critical community services to remain powered during outages, safeguarding lives and promoting stability (Solar Energy Industries Association [SEIA], 2023).

The economic benefits of rooftop solar further reinforce its role as a cornerstone of sustainable energy. From the creation of local jobs in solar installation and maintenance to the financial savings on energy bills, rooftop solar supports economic growth within communities. Federal and state incentives, such as tax credits and net metering, make rooftop solar increasingly accessible, empowering households across all income levels to invest in clean energy (EnergySage, 2022). Community solar initiatives expand this reach even further, offering renters and those without suitable rooftops the ability to participate in and benefit from shared solar projects. This model of renewable energy democratizes access to clean power, making it attainable for a wide demographic and promoting energy equity (NREL, 2021).

Advancements in solar technology continue to make rooftop systems more efficient, affordable, and adaptable, transforming rooftops into clean energy hubs and driving down installation costs. As efficiency improves and storage options expand, rooftop solar becomes an increasingly viable option for millions of Americans, positioning it as a sustainable solution that can scale to meet the country's ambitious carbon reduction goals. Each rooftop installation, each watt generated, and each community that adopts solar brings the nation closer to a decentralized, resilient energy future (SEIA, 2023).

With supportive policies, streamlined permitting, and proactive community engagement, rooftop solar has the potential to lead America into a new era where energy independence and environmental stewardship go hand in hand. Rooftop solar provides a model of renewable energy that not only benefits the environment but empowers people, paving the way for a future where clean energy is accessible, resilient, and rooted in the values of sustainability and community well-being. As the U.S. strives to reduce carbon emissions and build a sustainable energy infrastructure, rooftop solar stands out as a solution that aligns perfectly with these goals, offering a path to a greener, more equitable energy future that truly benefits people and the planet.

Chapter 15

Global Leaders in Alternative Energy and the Rise of Rooftop Solar

As the world moves toward a more sustainable energy future, the global race to expand alternative energy capacity has led certain nations to emerge as leaders in renewable energy adoption. Countries like China, the United States, Brazil, India, and Germany top the charts in terms of total installed renewable energy capacity. However, while large-scale projects like solar farms, wind farms, and hydroelectric plants have played a significant role in advancing clean energy in these nations, there is a growing interest in rooftop solar as a decentralized, efficient, and environmentally friendly solution. Among the global leaders in rooftop solar installations, Australia, Germany, and Japan have distinguished themselves with extensive per capita installations, setting the bar for accessible, community-centered energy solutions.

In contrast, although China and the United States rank first and second globally in terms of total renewable energy capacity, they lag considerably behind the global leaders in rooftop solar installations per capita. This chapter examines the renewable energy achievements of the top countries and explores the reasons why China and the U.S. have yet to fully capitalize on the benefits of rooftop solar compared to smaller but more dedicated nations.

Global Leaders in Renewable Energy Capacity: China, the United States, Brazil, India, and Germany

As of 2023, the countries leading the way in renewable energy capacity are:

1. **China**: China dominates the renewable energy sector with a staggering 1,453 gigawatts (GW) of installed capacity, largely due to massive investments in hydroelectric power, wind farms, and solar energy projects. Over the last decade, China has rapidly scaled up its clean energy infrastructure, driven by national policies aimed at reducing air pollution and meeting international climate commitments. Mega-projects like the Three Gorges Dam and vast solar farms in the Gobi Desert illustrate China's large-scale approach to renewable energy (Statista, 2023).

2. **United States**: With approximately 388 GW of renewable energy capacity, the United States holds the second position globally. Much of this capacity is driven by wind farms and large-scale solar projects, especially in states like Texas, California, and Iowa. U.S. federal and state-level incentives, alongside falling technology costs, have supported this growth, although centralized power generation remains a primary focus over decentralized, community-based solutions like rooftop solar (Statista, 2023).

3. **Brazil**: Brazil's renewable capacity reaches around 176 GW, with a heavy emphasis on hydroelectric power, accounting for over 60% of its energy production. However, Brazil has also seen growth in wind and solar installations as the country seeks to diversify its renewable energy sources.

4. **India**: India has approximately 168 GW of renewable capacity, with ambitious targets to reach 500 GW by 2030. Investments

in large-scale solar and wind projects, especially in states like Rajasthan and Gujarat, are central to India's renewable energy strategy.

5. **Germany**: Germany is a global leader in solar energy with around 147 GW of renewable energy capacity. Its commitment to renewable energy extends beyond large-scale installations; Germany is a strong supporter of community energy, decentralized power generation, and rooftop solar.

Leaders in Rooftop Solar: Australia, Germany, and Japan

While large-scale projects have defined the renewable energy landscape, rooftop solar is increasingly recognized for its accessibility and efficiency, especially in densely populated or urban areas. In terms of rooftop solar installations per capita, Australia, Germany, and Japan lead the way, driven by supportive policies, high electricity costs, and public demand for sustainable energy:

1. **Australia**: Australia is the undisputed global leader in rooftop solar, with an average of 746 watts (W DC) of rooftop solar capacity per person. With high sunlight availability and strong government incentives, Australia has made rooftop solar an accessible and attractive option for its residents. The high cost of traditional electricity has further incentivized Australian homeowners and businesses to adopt rooftop solar.

2. **Germany**: With approximately 668 W DC of rooftop solar per person, Germany has established a decentralized energy model that emphasizes both environmental protection and energy independence. Germany's strong net metering policies and financial incentives, such as the Renewable Energy Sources Act, have made rooftop solar widely accessible.

3. **Japan**: Japan ranks third globally in rooftop solar, with 353 W DC per capita. Following the 2011 Fukushima nuclear disaster,

Japan prioritized renewable energy, particularly rooftop solar, as a means of diversifying its energy sources and reducing dependency on nuclear power. Japan's Feed-in Tariff system has been instrumental in encouraging rooftop installations among residents and businesses.

Rooftop Solar in China and the United States: High Potential, Low Adoption

Despite their leadership in overall renewable energy capacity, both China and the United States lag significantly behind other nations in rooftop solar adoption. While both countries have invested heavily in renewable energy infrastructure, their focus has primarily been on large-scale solar farms and wind power installations, often in rural or desert areas. These large projects contribute substantial megawatt capacity to the national grid but come with limitations. By prioritizing centralized, high-capacity projects, both China and the U.S. have missed some of the advantages associated with rooftop solar, such as reduced transmission losses, improved energy resilience, and community-based energy independence.

In China, government incentives and development goals have historically driven the construction of expansive solar farms, particularly in western regions like Xinjiang and Inner Mongolia, where open land is available and sunlight is abundant. However, these centralized projects require extensive transmission infrastructure to transport electricity to densely populated urban areas in the east, where demand is highest. This focus has led to issues with energy wastage due to transmission losses and curtailment, where produced solar energy goes unused because it cannot be efficiently transmitted or stored (Li & Luo, 2021). Though China has recently started encouraging rooftop solar in urban areas through pilot programs and subsidies, centralized energy generation still dominates, limiting the growth of decentralized, residential solar solutions.

Similarly, the U.S. has centered much of its renewable energy strategy on large-scale projects, with significant solar installations in states like California, Texas, and Nevada. These projects benefit from economies of scale but also present challenges, including land use conflicts, habitat disruption, and the need for transmission lines to carry power over long distances. Moreover, regulatory challenges, resistance from utility companies, and complex permitting processes in some states have slowed the adoption of rooftop solar, especially compared to countries like Australia and Germany, which have nationwide policies supporting small-scale, decentralized solar (Carley & Konisky, 2020). In many parts of the U.S., utilities still hold monopolies on power generation and distribution, making it difficult for rooftop solar to gain a foothold in a highly centralized system.

The emphasis on centralized projects in both countries reflects a tendency to favor high-capacity installations that contribute immediately to national energy goals. However, by not fully embracing rooftop solar, both China and the U.S. miss out on a model that empowers individuals and communities to generate their own power. Rooftop solar provides additional benefits, such as reducing grid demand, enhancing resilience against power outages, and decreasing dependency on long-distance transmission. Expanding decentralized solar solutions would not only diversify the renewable energy mix in these countries but also align with the global trend toward distributed, sustainable energy models that prioritize environmental stewardship and community involvement.

Despite their impressive achievements in renewable energy capacity, both China and the United States struggle with the widespread adoption of rooftop solar. In each country, structural, regulatory, and geographic challenges have hindered rooftop solar from reaching its full potential, even as centralized, large-scale solar projects continue to dominate their energy landscapes. For China, the focus on centralized energy solutions, combined with urban planning constraints, has limited the expansion of rooftop solar in densely populated cities. In the United States, a mix of inconsistent policies, utility resistance, and complex permitting processes has created a fragmented environment for rooftop solar, allowing adoption in certain states but leaving substantial barriers nationwide:

1. **China**: With an extensive renewable energy infrastructure, China leads the world in total solar capacity but lags in rooftop solar per capita. While China has made some progress in promoting rooftop solar in urban areas, particularly through initiatives in cities like Beijing and Shanghai, the centralized nature of China's energy sector has led to a preference for large-scale projects. Policy support for rooftop solar has been inconsistent, and urban planning challenges have limited the potential for widespread rooftop installations (Li & Luo, 2021). Additionally, as China's urban areas are densely populated and often high-rise, rooftop availability is limited, posing a challenge to scaling rooftop solar.

2. **United States**: In the U.S., rooftop solar adoption has gained traction in states like California, Hawaii, and Arizona, where incentives and favorable sunlight conditions make it a viable option. However, rooftop solar per capita still lags compared to Australia and Germany. Complex permitting processes, limited federal incentives, and utility companies' resistance to net metering in some states have created barriers to widespread rooftop solar adoption (Carley & Konisky, 2020). While some cities and states have made notable progress, the U.S. lacks a cohesive national policy that prioritizes rooftop solar, resulting in a fragmented landscape that limits broader adoption.

The Benefits of Rooftop Solar: Decentralized, Efficient, and Environmentally Friendly

The potential benefits of rooftop solar are immense, both for individual consumers and for the broader energy infrastructure. By generating electricity directly at the point of consumption, rooftop solar significantly reduces energy transmission losses, which can be as high as 5-10% in traditional, centralized grid systems due to the need for electricity to travel over long distances (National Renewable Energy Laboratory [NREL], 2021). This localized production allows for maximum utilization of each kilowatt generated, translating into greater efficiency and cost savings for consumers. Additionally, rooftop solar alleviates the environmental impacts associated with large-scale land use in expansive renewable projects, such as habitat disruption and the clearing of natural landscapes. By leveraging existing building structures—residential rooftops, commercial spaces, and public buildings—rooftop solar minimizes the need for new land, making it a sustainable choice for energy expansion, particularly in urban areas.

Countries leading in rooftop solar adoption, such as Australia and Germany, showcase the advantages of a decentralized energy model. This approach not only enhances energy independence by reducing reliance on centralized grids but also creates more resilient local communities that are less vulnerable to large-scale grid failures. Decentralization diversifies the power supply, making communities less dependent on remote power plants and transmission infrastructure that may be vulnerable to weather disruptions, technical failures, or geopolitical issues (EnergySage, 2022). Rooftop solar enables homeowners and communities to gain a degree of energy self-sufficiency, empowering them to meet their own energy needs sustainably while contributing to a reduced carbon footprint.

Rooftop solar aligns particularly well with urban environments, where rooftops present a readily available resource for renewable energy generation. In densely populated cities, rooftop installations can help offset high demand for electricity, relieving pressure on local grids during peak times. Urban rooftops offer an often-underutilized space that, when transformed into solar power hubs, can support sustainable energy generation without infringing on natural or agricultural lands. In cities worldwide, solar panel installations are increasingly being integrated into rooftop designs, turning buildings into active participants in clean energy production while lowering greenhouse gas emissions at the local level (Solar Energy Industries Association [SEIA], 2023).

Beyond efficiency and environmental benefits, rooftop solar systems are particularly valuable for creating resilient energy communities. Because they generate power locally, rooftop systems can continue to function independently of the main grid during outages, providing critical backup power to homes, businesses, and essential services. This resilience is particularly advantageous for countries like the U.S. and China, where much of the infrastructure development has focused on centralized, large-scale projects. By shifting focus toward rooftop solar, these countries could not only improve grid stability but also make significant strides in energy accessibility and resilience. Rooftop solar helps to disperse energy resources, reducing the strain on transmission lines and enabling a more stable, flexible energy grid that can adapt to both growing demand and environmental challenges.

Furthermore, increasing rooftop solar capacity supports global sustainability goals by reducing emissions, conserving land, and promoting a low-carbon economy. For countries with ambitious renewable energy targets, rooftop solar offers a pathway that is compatible with both economic and environmental objectives. Unlike large-scale projects, which may be hampered by geographic constraints or land-use conflicts, rooftop solar can be installed incrementally and scaled to meet local needs. This adaptability makes rooftop solar a vital component of a sustainable energy strategy, aligning with global goals to mitigate climate change and transition to clean, renewable energy sources (NREL, 2021).

By shifting more attention and resources toward rooftop solar, countries can create more resilient, efficient, and environmentally conscious energy systems. Rooftop solar offers a unique combination of benefits—reduced transmission losses, environmental conservation, enhanced energy independence, and community resilience—that make it a foundational element of a decentralized, sustainable energy future.

Unlocking the Potential of Rooftop Solar in China and the United States

For China and the United States to realize the full potential of rooftop solar, policy changes and public support are essential. Simplified permitting, tax incentives, and stronger net metering policies could encourage wider adoption of rooftop solar in both countries. As rooftop solar technology becomes more efficient and affordable, a shift toward decentralized energy production could foster energy independence, reduce costs, and support environmental stewardship. Increasing rooftop solar adoption would not only help China and the U.S. meet their ambitious renewable energy goals but also position them as leaders in sustainable, community-centered energy models.

By learning from countries like Australia and Germany, which have successfully implemented policies to promote rooftop solar, China and the United States can build more resilient and sustainable energy systems that empower individuals, protect natural landscapes, and reduce dependency on centralized power plants

Chapter 16

Rooftop Solar is the Better Investment for America's Future

The future of America's energy system is at a critical crossroads, and the decisions made today will have far-reaching consequences for generations to come. While large-scale alternative energy projects have been hailed as the solution to the nation's energy and environmental challenges, their track record suggests otherwise. In contrast, rooftop solar emerges as a superior option that is not only more efficient and cost-effective but also better aligned with environmental sustainability and energy independence. As the nation moves forward, it is essential that policymakers, businesses, and individuals recognize the benefits of decentralized energy systems like rooftop solar and commit to supporting their widespread adoption.

Summarizing the Failure of Large-Scale Alternative Energy Projects

Large-scale renewable energy projects, such as utility-scale solar farms, wind farms, and massive hydroelectric dams, were initially seen as the ideal solution to the country's growing energy demands. However, these projects have faced several significant challenges that have limited their success and effectiveness, such as:

1. **Land Use and Environmental Impact**: Large-scale renewable energy projects often require vast tracts of land, leading to land degradation, habitat destruction, and disruptions to local ecosystems. For example, utility-scale solar farms in the desert have been criticized for their impact on fragile ecosystems and wildlife. Similarly, wind farms can pose threats to bird and bat populations, and hydroelectric dams disrupt aquatic ecosystems and fish migration patterns. These unintended consequences undermine the environmental benefits that these projects were intended to deliver (Hernandez et al., 2014).

2. **Transmission and Distribution Challenges**: Large-scale energy projects are typically located in remote areas, far from population centers where the energy is most needed. This requires the construction of extensive transmission lines, which adds significant costs and leads to energy losses during transmission. Additionally, these projects are vulnerable to natural disasters and grid failures, making them less reliable as energy sources. The costs associated with maintaining and upgrading the transmission infrastructure can also be prohibitive, slowing down the deployment of renewable energy on a large scale (Borenstein, 2017).

3. **High Upfront Costs and Long Development Timelines**: The development of large-scale renewable energy projects often involves significant upfront capital investment and long lead times, making them less attractive to investors. These projects can take years to plan, permit, and construct, delaying their ability to provide clean energy to the grid. Moreover, the costs associated with environmental assessments, regulatory compliance, and securing land rights further complicate the development process, reducing the cost-efficiency of these projects (Millstein et al., 2017).

Large Scale Energy Alternative Energy Projects

While large-scale projects have undoubtedly contributed to the growth of renewable energy in the U.S., their inherent limitations highlight that they cannot fully meet the country's alternative energy needs on their own. These expansive facilities, whether solar or wind farms, often demand vast areas of land, leading to the clearing of habitats and, in many cases, the disruption of sensitive ecosystems. Placed in remote locations to capitalize on optimal wind or sunlight, they require extensive transmission infrastructure to deliver power to distant urban centers, resulting in energy losses along the way. This sprawling infrastructure comes at an environmental cost, with potential damage to landscapes, wildlife corridors, and local biodiversity. Furthermore, large-scale projects are frequently dependent on government subsidies and corporate investment, shifting the focus from genuine sustainability toward profit-driven motives that may overlook long-term ecological impacts. Compounding these issues, the intermittent nature of centralized solar and wind facilities necessitates substantial backup resources—often natural gas—to ensure grid stability, which reduces their overall environmental benefits and complicates efforts to decarbonize.

In contrast, rooftop solar and other decentralized renewable solutions provide a more adaptable, efficient, and environmentally friendly alternative. By generating power directly at the point of consumption, rooftop solar effectively minimizes transmission losses and reduces the need for extensive infrastructure, making productive use of already-developed spaces like rooftops, parking structures, and other urban spaces. This approach alleviates the strain on open lands and reduces the ecological footprint associated with renewable energy production. Decentralized energy solutions also empower individuals, businesses, and communities to take control of their energy needs, fostering a sense of ownership and promoting energy independence. By enabling local energy production, rooftop solar enhances resilience against grid disruptions, particularly valuable in areas prone to extreme weather events where centralized grids are vulnerable.

Decentralized energy offers economic benefits as well, stimulating local economies by creating jobs in solar installation, maintenance, and energy management. Unlike large-scale projects that may concentrate profits within corporations, local renewable solutions distribute financial benefits more evenly, supporting local businesses and providing homeowners with long-term savings on their energy bills.

Additionally, community solar initiatives can extend these benefits to renters and those unable to install solar on their own properties, promoting inclusivity and allowing a wider demographic to participate in the shift toward clean energy.

As the nation aims to reduce carbon emissions and bolster energy security, a shift toward smaller, community-based renewable solutions like rooftop solar is essential. These alternatives present a pathway to a cleaner, more sustainable energy future—one that aligns more closely with environmental stewardship, reduces dependency on corporate-driven projects, and prioritizes the needs and values of local communities. By emphasizing decentralized energy, the U.S. can build an energy landscape that supports environmental goals, empowers communities, and establishes a more resilient, distributed energy infrastructure.

Rooftop Solar as the More Efficient, Cost-Effective, and Environmentally Friendly Option

In contrast to the numerous challenges associated with large-scale renewable energy projects, rooftop solar provides a range of benefits that make it a more viable and sustainable investment for America's future. Large-scale solar farms and wind projects often require vast amounts of land, disrupt natural habitats, and demand extensive infrastructure to transmit energy to urban centers. These factors not only drive up costs but also result in significant energy losses during transmission, as well as lasting impacts on local ecosystems.

Rooftop solar, on the other hand, offers a localized approach to energy production, generating electricity directly at the point of consumption. This eliminates the need for costly, long-distance transmission and minimizes energy losses, making rooftop solar inherently more efficient. By utilizing already-developed spaces, such as residential and commercial rooftops, parking structures, and even public buildings, rooftop solar optimizes land use without encroaching on natural habitats or farmland. This not only preserves ecosystems but also supports sustainable land management by reducing the need to clear open spaces for renewable infrastructure.

Financially, rooftop solar provides significant long-term savings for both individual homeowners and communities. Federal and state incentives, including tax credits and rebates, help reduce the initial cost of installation, making solar more accessible to a broader range of households. Over time, homeowners see substantial reductions in their energy bills, with many systems paying for themselves within five to ten years. Additionally, programs like net metering allow homeowners to earn credits for any excess energy they generate, further increasing their return on investment. Unlike large-scale projects, which often rely on ongoing subsidies and corporate funding, rooftop solar creates financial independence, empowering individuals and communities to take control of their energy costs.

Environmentally, rooftop solar is a less intrusive alternative that aligns closely with conservation goals. By generating power close to where it's consumed, rooftop solar avoids the need for sprawling infrastructure, reducing both the ecological and visual impact associated with traditional energy production. Rooftop solar also supports the development of more resilient, decentralized energy systems, which are essential in the face of climate-related disruptions. Localized energy production ensures that communities have access to power even if centralized grids are compromised by extreme weather events, enhancing overall energy resilience.

The environmental, financial, and efficiency benefits of rooftop solar position it as a forward-looking solution for sustainable energy. By embracing this approach, the U.S. can move toward a cleaner, more cost-effective energy landscape that empowers communities, minimizes environmental impact, and reduces dependency on massive, corporate-driven projects. Rooftop solar isn't just a green choice; it's a practical and impactful strategy for building a resilient energy future aligned with America's sustainability goals. Here are the benefits of rooftop solar:

1. **Efficiency and Cost-Effectiveness**: Rooftop solar systems are installed directly on homes, businesses, and public buildings, allowing for energy generation at the point of consumption. This eliminates the need for costly transmission infrastructure and minimizes energy losses during distribution. The cost of installing rooftop solar has dropped significantly in recent years, thanks to advances in solar technology, economies of scale, and supportive government policies such

as the Investment Tax Credit (ITC). Homeowners and businesses that invest in rooftop solar can see immediate reductions in their electricity bills, with the potential to recoup their investment in as little as five to seven years (Seel, Barbose, & Wiser, 2014).

Moreover, rooftop solar systems can be combined with battery storage to provide even greater savings and resilience. Energy generated during the day can be stored for use at night or during periods of high demand, reducing reliance on the grid and lowering costs even further. In contrast to large-scale projects that require massive capital investments and long development timelines, rooftop solar installations can be completed in a matter of weeks, providing a much faster return on investment.

2. **Environmental Benefits**: Rooftop solar is inherently more environmentally friendly than large-scale projects. It makes use of existing structures, such as roofs, rather than requiring the development of new land. This minimizes the environmental footprint of energy production and avoids the ecological disruptions associated with large-scale renewable projects. By generating clean, renewable energy directly where it is needed, rooftop solar reduces the need for fossil fuels, cutting greenhouse gas emissions and improving air quality. Additionally, because rooftop solar systems are decentralized, they enhance the resilience of local communities by providing a reliable source of energy during grid outages or natural disasters (Millstein et al., 2017).

3. **Energy Independence and Decentralization**: One of the most significant advantages of rooftop solar is its role in promoting energy independence. Instead of relying on large utility companies or remote power plants, homeowners and businesses can generate their own electricity, reducing their vulnerability to price fluctuations and energy shortages. This decentralization of energy production not only empowers individuals but also creates a more resilient energy system overall. In times of crisis, such as extreme weather events or

cyberattacks, decentralized energy systems are less likely to experience widespread outages, as they are not dependent on a single, centralized source of power (Owens & Faruqui, 2011).

4. **Scalability and Accessibility**: Rooftop solar offers greater scalability than large-scale renewable energy projects. While the development of utility-scale solar farms or wind farms can be delayed by land acquisition, permitting, and financing challenges, rooftop solar systems can be installed incrementally, one building at a time. This makes rooftop solar accessible to a wide range of consumers, from individual homeowners to small businesses and large corporations. Community solar programs and innovative financing options, such as solar leases and power purchase agreements (PPAs), have made it easier for low- and middle-income households to participate in the clean energy transition, further democratizing access to renewable energy (Lazar & Gonzalez, 2015).

A Call to Action for Policymakers, Businesses, and Individuals to Embrace Decentralized Energy

To fully realize the benefits of rooftop solar and decentralized energy, a concerted effort is needed from policymakers, businesses, and individuals. Policymakers must prioritize policies that support the growth of decentralized energy systems by enacting and maintaining tax incentives, strengthening net metering programs, and streamlining permitting processes for solar installations. At the federal level, the extension of the Investment Tax Credit (ITC) for solar energy is crucial to maintaining the momentum of rooftop solar adoption. States should also implement policies that encourage the development of community solar programs, allowing renters and those without suitable roofs to participate in the clean energy transition.

Utility companies must also adapt to the decentralized energy future. Rather than viewing rooftop solar as a threat to their business models, utilities should embrace the role of energy managers and facilitators. This includes investing in smart grid technologies, which allow for the integration of distributed energy resources, and offering energy storage and management solutions to consumers. By doing so, utilities can position themselves as essential players in the new energy landscape, creating new revenue streams while supporting the transition to clean energy.

Finally, individuals and businesses must recognize the benefits of investing in rooftop solar. For homeowners, rooftop solar provides long-term savings, energy independence, and environmental benefits. For businesses, it offers a way to reduce operational costs, enhance corporate social responsibility, and insulate against rising energy prices. By embracing rooftop solar, individuals and businesses can play a critical role in driving the clean energy revolution.

The evidence is undeniable: rooftop solar is the most efficient, cost-effective, and environmentally responsible investment for America's energy future. Unlike large-scale alternative energy projects, which are often granted for the benefit of corporate interests rather than the public good, rooftop solar empowers individuals and communities by reducing energy costs and fostering energy independence. It offers an energy model that is not only more affordable and sustainable but also has the least impact on the wider environment. While pork-barrel mega-projects consume vast tracts of land, disrupt ecosystems, and require billions in taxpayer funding, rooftop solar makes use of existing structures—homes, businesses, and public buildings—without the need for additional land or massive infrastructure development.

The time has come to end these large, inefficient projects that serve corporate America at the expense of the taxpayer. Politicians must stop doing favors for their campaign contributors by greenlighting expensive mega-projects that do little to benefit the average citizen. These corporate-driven, politically-motivated projects only serve to enrich a few while placing the burden of cost and environmental degradation on the public. Rooftop solar, by contrast, is a decentralized, democratized solution that puts power—both figuratively and literally—back into the hands of the people.

To ensure a truly sustainable and equitable energy future, policymakers, businesses, and individuals must come together to support the widespread adoption of decentralized energy systems like rooftop solar. The path forward is clear: we must shift our focus away from wasteful, large-scale projects designed to benefit corporations and embrace a smarter, more responsible energy strategy. By taking these necessary steps today, we can create a cleaner, more resilient, and more prosperous future for all Americans—one where the benefits of renewable energy are shared by everyone, not just the privileged few.

In doing so, we can build an energy system that not only reduces our carbon footprint but also breaks the cycle of political favoritism and corporate greed that has long dominated America's energy landscape. Rooftop solar is the key to unlocking this future, and it's up to us—policymakers, business leaders, and citizens (i.e. *the tax payers*)—to make this vision a reality. The choices we make today will define the future—we must act now to build a cleaner, more resilient, and prosperous tomorrow for all Americans.

Reference

American Wind Energy Association. (2021). Land use impacts of wind farms in Texas. Retrieved from https://www.awea.org

Baker, L. (2019). Distributed Energy and the Future of Utilities. Routledge.

Baker, S. (2022). The case for rooftop solar over large-scale energy farms. Energy Policy Journal, 128(1), 24-32. https://doi.org/10.1016/j.enpol.2022.128032

Bird, L., Reger, A., Heeter, J., & O'Shaughnessy, E. (2018). Policies and Market Trends Supporting Distributed Solar. National Renewable Energy Laboratory.

Borenstein, S. (2005). Time-varying retail electricity prices. Energy Economics, 27(2), 285-306.

Borenstein, S. (2017). The Economics of Distributed Solar Power. Energy Institute at Haas.

BrightSource Energy. (2021). The Ivanpah solar electric generating system. Retrieved from https://www.brightsourceenergy.com/ivanpah

California Energy Commission. (2023). Ivanpah Solar Electric Generating System: Energy output and environmental impact report. Retrieved from https://www.energy.ca.gov

Carley, S., & Konisky, D. M. (2020). The challenges of transitioning to a renewable energy economy in the United States. Annual Review of Environment and Resources, 45, 113 134. https://doi.org/10.1146/annurev-environ-012320-083401

Carolan, M. (2018). Land use conflicts in wind farm development: A socio-environmental perspective. Rural Sociology, 83(2), 278-299.

Cohen, J. (2019). Renewable energy policy and the future of electricity grids. Energy and Environment, 30(4), 559-572. https://doi.org/10.1177/0958305X19865210

Darghouth, N. R., Wiser, R., Barbose, G., & Mills, A. (2016). Net metering and market feedback loops: Exploring the impact of retail rate design on distributed PV deployment. The Energy Journal, 37(4).

Defenders of Wildlife. (2022). The Mojave Desert tortoise: Habitat and conservation efforts. Retrieved from https://www.defenders.org

Denholm, P., O'Connell, M., Brinkman, G., & Jorgenson, J. (2021). The environmental footprint of utility-scale solar projects. Renewable Energy, 92, 44-53.

EnergySage. (2021). How much power does a typical solar panel system produce? Retrieved from https://www.energysage.com

EnergySage. (2022). The impact of technological advancements on rooftop solar efficiency and cost. Retrieved from https://www.energysage.com

Energy Information Administration (EIA). (2021). How much electricity is lost in transmission and distribution in the United States?. Retrieved from https://www.eia.gov/tools/faqs/faq.php?id=105&t=3

Felder, F. A., & Athawale, R. (2014). The life and death of the utility death spiral. The Electricity Journal, 27(6),

Hernandez, R. R., Hoffacker, M. K., & Field, C. B. (2014). Land-Use Efficiency of Big Solar. Environmental Science & Technology, 48(2), 1315-1323.

Hernandez, R. R., Easter, S. B., Murphy-Mariscal, M. L., Maestre, F. T., Tavassoli, M., Allen, E. B., & Allen, M. F. (2015). Environmental impacts of utility-scale solar energy. Renewable and Sustainable Energy Reviews, 29, 766-779.

Hogan, W. W. (2014). Electricity Market Design: Time-of-Use Pricing and Smart Grid Integration. Journal of Regulatory Economics, 45(2), 128-149.

International Renewable Energy Agency (IRENA). (2020). Solar PV technology brief. https://www.irena.org

Joskow, P. L. (2019). Challenges for wholesale electricity markets with intermittent renewable generation at scale: The US experience. Oxford Review of Economic Policy, 35(2), 291-331. https://doi.org/10.1093/oxrep/grz010

Lazard. (2019). Levelized Cost of Energy Analysis—Version 13.0. Lazard.

Lazar, J., & Gonzalez, W. (2015). Smart Rate Design for a Smart Future. Regulatory Assistance Project.

Li, J., & Luo, Y. (2021). Decentralized solar adoption in China: Barriers and potential. Renewable Energy Development Journal, 22(3), 234-247.

Lovich, J. E., & Ennen, J. R. (2017). Environmental impacts of solar and wind power in fragile ecosystems. Environmental Management Journal, 60(1), 104-112.

Lovich, J. E., & Ennen, J. R. (2017). Desert tortoises and renewable energy development: Ensuring conservation while promoting clean energy. BioScience, 63(12), 993-998.

Millstein, D., Wiser, R., Bolinger, M., & Barbose, G. (2017). The climate and air-quality benefits of wind and solar power in the United States. Nature Energy, 2(9), 17134.

National Renewable Energy Laboratory (NREL). (2020). Ivanpah Solar Electric Generating System: A case study in land use impacts. Retrieved from https://www.nrel.gov

National Renewable Energy Laboratory (NREL). (2021). Solar energy and the economics of rooftop systems. https://www.nrel.gov

National Renewable Energy Laboratory (NREL). (2021). Energy loss during transmission in centralized vs. decentralized systems. Retrieved from https://www.nrel.gov

National Renewable Energy Laboratory (NREL). (2022). Rooftop solar power potential in the United States. Retrieved from https://www.nrel.gov

Owens, B., & Faruqui, A. (2011). The Rise of Distributed Energy and the Role of Utilities. The Electricity Journal, 24(4), 52-60.

Pasqualetti, M. J. (2019). The land use dilemma of renewable energy development: Impacts on ecosystems and economies. Renewable and Sustainable Energy Reviews, 113, 109290.

Seel, J., Barbose, G., & Wiser, R. (2014). An analysis of residential PV system price differences between the United States and Germany. Energy Policy, 69, 216-226.

Solar Energy Industries Association. (2021). The benefits of rooftop solar for sustainable energy. Retrieved from https://www.seia.org

Solar Energy Industries Association. (2023). Federal solar tax credit guide. https://www.seia.org

Sovacool, B. K. (2017). The political economy of energy transitions: Wind and solar power in the Tomaschek, A., & Miller, R. (2021). The environmental impact of large-scale renewable energy projects: A critical review. Journal of Environmental Management, 285, 112-124.

U.S. Department of Energy. (2016). Energy Policy Act of 2005: 10 years later. https://www.energy.gov/sites/prod/files/2016/03/f30/Energy-Policy-Act-of-2005.pdf

U.S. Department of Energy. (2019). Wind power in Texas: Transmission challenges and solutions. https://www.energy.gov/texaswindreport

U.S. Department of Energy. (2020). Crescent Dunes Solar Energy Project: An analysis of failure and lessons learned. Retrieved from https://www.energy.gov

U.S. Department of Energy. (2022). Challenges of concentrated solar power (CSP). Retrieved from https://www.energy.gov

U.S. Energy Research & Social Science, 28, 5-15. https://doi.org/10.1016/j.erss.2017.03.020

U.S. Environmental Protection Agency (EPA). (2022). Air quality and health impacts of fossil fuel emissions. Retrieved from https://www.epa.gov

Wang, T., & Wang, P. (2020). Challenges of integrating renewable energy into the grid: The case of large-scale wind farms. Renewable Energy Review, 144, 1120-1129. https://doi.org/10.1016/j.rser.2020.102423

Warren, C. R., & Birnie, R. V. (2020). Revolt of the locals: Wind farms, local opposition, and planning dilemmas. Environmental Planning A, 34(8), 285-299.

About the Author

Dr. Douglas B. Sims is an environmental expert with over 30 years of experience in the consulting industry, specializing in large-scale alternative energy projects and NEPA (National Environmental Policy Act) permitting. His career spans projects from 100-acre solar fields to 500,000-acre wind energy developments, where he has provided critical expertise on environmental impacts, land use, and sustainable planning.

Sims' skill in navigating regulatory frameworks has been key to the success of many large-scale developments, helping clients meet environmental standards while advancing renewable energy. His work carefully balances innovation and ecological responsibility, ensuring projects protect ecosystems and promote sustainability.

In addition to consulting, Dr. Sims has published papers on soil remediation and environmental contamination in peer-reviewed journals. His academic and practical experience have made him a sought-after consultant, particularly in projects requiring expertise in environmental geochemistry and contamination. Outside of work, he is passionate about the intersection of environmental science, politics, and societal issues, offering insights into public policy and energy innovation.

Married to his college sweetheart since the mid-1990s, Dr. Sims and his wife have raised two children and continue to share their deep commitment to education, environmental responsibility, and curiosity about the world.

www.ingramcontent.com/pod-product-compliance
Lightning Source LLC
Chambersburg PA
CBHW071425210326
41597CB00020B/3659
* 9 7 9 8 9 9 1 9 1 0 8 2 8 *